CONTENTS

CHRONOLOGICAL LIST OF HEARINGS

2003

THURSDAY, OCTOBER 30, 2003

DESTRUCTION OF THE U.S. CHEMICAL WEAPONS STOCKPILE—PROGRAM STATUS AND ISSUES

STATEMENTS PRESENTED BY MEMBERS OF CONGRESS

WITNESSES

APPENDIX

DESTRUCTION OF THE U.S. CHEMICAL WEAPONS STOCKPILE—PROGRAM STATUS AND ISSUES

HOUSE OF REPRESENTATIVES,
COMMITTEE ON ARMED SERVICES,
TERRORISM, UNCONVENTIONAL THREATS AND CAPABILITIES SUBCOMMITTEE,
Washington, DC, Thursday, October 30, 2003.

The subcommittee met, pursuant to call, at 9:03 a.m., in room 2118, Rayburn House Office Building, Hon. Jim Saxton (chairman of the subcommittee) presiding.

OPENING STATEMENT OF HON. JIM SAXTON, A REPRESENTATIVE FROM NEW JERSEY, CHAIRMAN, TERRORISM, UNCONVENTIONAL THREATS AND CAPABILITIES SUBCOMMITTEE

Mr. SAXTON. Good morning.

Today, the Terrorism, Unconventional Threats and Capabilities Subcommittee meets to review the Department of Defense (DOD) program for destruction of the U.S. stockpile of lethal chemical warfare agents and munitions. We are joined in the hearing by several Members of Congress who have chemical stockpile storage sites in their districts and who are interested in the chemical demilitarization program. I want to welcome those gentlemen.

The U.S. chemical weapons stockpile originally consisted of approximately 31,000 tons of lethal chemical agents in a wide variety of munitions, located at Johnston Atoll in the Pacific southwest of Hawaii and eight sites in the continental United States.

The fiscal year 1986 Defense Authorization Act requires that the destruction of the stockpile be carried out so as to ensure maximum protection of the environment, the general public and the workers at the storage and demilitarization sites. Destruction of the stockpile began at Johnston Atoll in 1990 and is supposed to be completed by April 29, 2007, in accordance with the Chemical Weapons Convention Treaty.

As of September 24, 2003, approximately 8,220 tons—or 26 percent—of the stockpile has been destroyed. Our witnesses today will talk about progress in the chemical demilitarization program, but will also talk about issues related to program execution, management, cost and schedule.

As members of congress, we will hear good news and bad news. The good news will the progress being made in the program as new chemical weapons destruction facilities are brought on line, weapons are destroyed and safety of the public and the environment is increased as the stockpile is reduced.

The bad news will be estimates of increased time to complete the destruction program and the increased program costs and implications for commitments under the Chemical Weapons Convention Treaty that will result.

The total cost of the program has grown from an estimated $1.7 billion in 1986 when the program was initiated at the direction of Congress to an estimated $25 billion today. The estimated time to complete destruction of the stockpile now extends into the next decade.

A number of key factors have affected the ability of the Department of Defense and the Army to effectively control the cost and schedule of the chemical stockpile destruction program, many of which are not under the control of either organization. The chemical demilitarization program is very large and complex, and is influenced by a number of offices and entities within and outside the Department of Defense, not the least of which has been Congress.

The issue for Congress and the subcommittee today is to gain an understanding of the progress being made in the chemical weapons destruction program, factors affecting the growth of the program schedule and cost and what might be done to accelerate the completion of the program, both from the standpoint of reducing the cost of the program and ensuring the maximum protection of the public, the personnel involved in destruction of the stockpile and the environment.

Our witnesses today include: Mr. Henry L. Hinton, Jr., from the General Accounting Office; Mr. Patrick Wakefield, Deputy Assistant to the Secretary of Defense in the area of chemical demilitarization and counterproliferation; the Honorable Claude M. Bolton, Assistant Secretary of the Army for Acquisitions, Logistics and Technology; Mr. Michael A. Parker, Director of the U.S. Army Chemical Materials Agency; and Mr. Craig Conklin from the Federal Emergency Management Agency.

We want to welcome you gentlemen. And we look forward to your testimony.

But before we begin our witnesses, let me turn to our great ranking member, Mr. Meehan, for his opening statement.

[The prepared statement of Mr. Saxton can be found in the Appendix on page 35.]

STATEMENT OF HON. MARTIN T. MEEHAN, A REPRESENTATIVE FROM MASSACHUSETTS, RANKING MEMBER, TERRORISM, UNCONVENTIONAL THREATS AND CAPABILITIES SUBCOMMITTEE

Mr. MEEHAN. Thank you very much, Mr. Chairman. And I am pleased to be here this morning and to welcome our panelists.

From where I sit, there is much to say in praise of the ChemDemil program. For example, at Johnston Atoll and Tooele, we have successfully demonstrated incineration technology and destroyed more than 26 percent of the chemical stockpile without the loss of even one single human life. Further, through aggressive exploratory research and development efforts, we have promoted and accepted alternative disposal technologies in response to various environmental concerns. Yet, these highlights cannot receive praise without the mention of many of the downsides of the program.

First, while positive in some light, the alternative technologies have inadvertently contributed to the effort's overall cost growth, a definite downside indeed, as the cost projections are now well above $25 billion. Second, the destruction schedule has slipped and is not expected to reach completion until the year 2014, more than a decade beyond initial projections. Third, while it is true that the 26 percent of the stockpile has been destroyed, more than 23,000 tons of material remain, providing perhaps a potentially rich target for terrorists seeking to wreak havoc on American citizens. And fourth, the current schedule fails to comply with the deadlines set forth in the Chemical Weapons Conventions Treaty.

While I applaud the programmatic highlights and accomplishments, I remain concerned about the setbacks and persistent challenges. While I recognize that the Russians also are not on schedule to comply with the treaty, and while I also understand that much of the cost growth is simply the result of inflation, I feel compelled to mention that politics, indecisive management and a lack of true commitment to funding has led to the current state with which we now find ourselves.

I am frustrated that other DOD priorities have repeatedly won out in the resource game. And I am frustrated that the Army has transferred program management from one office to the next. And I am frustrated with the political opportunism, both within the Pentagon and throughout our nation's communities, and recognize that its fallout has led to one delay after another, as we attempt to erase this residue of the Cold War and eliminate a possible Achilles heel on our domestic security.

Mr. Chairman, I am eager to listen to today's testimony. I hope that many of my concerns can be addressed. And I stand ready to work with the department to advance this effort in a cost-effective, environmentally friendly manner and hope that today's meeting will shed light on opportunities that may be available in this regard.

Thank you again, Mr. Chairman, for calling the panel's attention to this important issue. And I yield back my time.

Mr. SAXTON. I thank the ranking member for a very fine statement.

Mr. Hinton, you will be our first witness. Before you begin, I would like to ask unanimous consent to enter into the record a report by the Department of Defense, Office of Inspector General, entitled, "The Chemical Demilitarization Program: Increased Costs for Stockpile and Non-Stockpile Chemical Material Disposal Programs;" actually, report number D–2003–128, dated September 4, 2003.

The report discusses the factors that continue to affect the cost and schedule of the chemical stockpile program. And copies of the report are available on the inspector general's website.

Hearing no objection, Mr. Hinton, please, the floor is yours, sir.

[The information referred to can be found in the Appendix on page 81.]

STATEMENT OF HENRY HINTON, JR., MANAGING DIRECTOR, DEFENSE CAPABILITIES AND MANAGEMENT, GENERAL ACCOUNTING OFFICE

Mr. HINTON. Thank you, Mr. Chairman, Mr. Meehan.

I am pleased to be here this morning to testify on DOD's program to destroy the nation's chemical weapons stockpile. My testimony today addresses four issues, mainly from General Accounting Office (GAO) September 2003 report, in which we concluded: "Sustained leadership, along with key strategic management tools, is needed to guide DOD's destruction program."

That report was requested by the chairmen of the House and Senate Armed Services Committees. First, the program has had many schedule delays and its costs keep rising.

In 2001, the program extended the destruction schedule and increased the total cost estimate from $15 billion to about $24 billion. Since then, nearly all sites have experienced delays, stemming from problems such as: plant safety issues; compliance with environmental requirements; unresolved emergency preparedness issues; and funding shortfalls.

Program officials say the delays have raised the cost estimate by an additional $1.4 billion to more than $25 billion today. Based on current schedule slippages, GAO believes that further delays will occur and costs will grow even higher. The program management team needs to develop a risk management plan to mitigate the problems affecting the program schedules, cost and safety issues.

Second, because of schedule delays, the United States will not meet the Chemical Weapons Convention's (CWC) deadline of April 2004 to destroy 45 percent of the stockpile. It also risks not meeting the original 2007 deadline to complete destruction of the entire stockpile. Unless the program fixes the problems causing the delays, the United States also risks not meeting the convention's deadline of 2012 if it is extended.

Third, the program has suffered from several longstanding management and organizational issues. The lack of sustained leadership has undercut decisionmaking authority and obscured accountability.

The current program's complex structure, with multiple lines of authority, has left roles and responsibilities unclear. DOD does not have an overarching, comprehensive strategy to guide and integrate its activity and monitor performance, as leading organizations embrace. Without these key elements, Mr. Chairman, DOD and the Army have no assurances of meeting their goal to destroy the chemical weapons stockpile in a safe and timely manner and within cost.

Fourth, some good news. The program has improved emergency preparedness in communities near the destruction sites. The Army and Federal Emergency Management Agency (FEMA) have helped state and local communities become better prepared to respond to chemical emergencies. However, like the rest of the program, costs of the emergency preparedness in the communities are rising. Some of the states with stockpiles have increased the amounts they are requesting beyond their approved budgets. These requests amount to about $88 million for fiscal years 2004 and 2005.

In conclusion, Mr. Chairman, GAO recommended in its report that the under secretary for defense for acquisition, technology and logistics, in conjunction with the Secretary of the Army, take two actions. One, develop an overall strategy and implementation plan for the program that would: articulate a program mission statement; identify the program's long-term goals and objectives; three, delineate the roles and responsibilities of all DOD and Army offices that are involved; and establish near-term performance measures.

Our second recommendation directed them to implement a risk management approach that anticipates and influences internal and external factors that could adversely impact program performance. DOD concurred with our recommendations and is taking steps to implement them. Mr. Chairman, that concludes my remarks. I stand ready to address yours and the committee's questions.

[The prepared statement of Mr. Hinton can be found in the Appendix on page 41.]

Mr. SAXTON. Thank you very much, sir.

Mr. Wakefield.

STATEMENT OF PATRICK WAKEFIELD, DEPUTY ASSISTANT TO THE SECRETARY OF DEFENSE (CHEMICAL DEMILITARIZATION AND COUNTERPROLIFERATION)

Mr. WAKEFIELD. Thank you, Mr. Chairman.

Mr. Chairman and distinguished committee members, I wish to thank you for the opportunity to appear before this committee today to discuss the United States Chemical Demilitarization Program. I am Patrick Wakefield, the Deputy Assistant to the Secretary of Defense for Chemical Demilitarization and Threat Reduction.

I am the single focal point within the Office of the Secretary of Defense responsible for the oversight, coordination and integration of the chemical demilitarization program, the Assembled Chemical Weapons Alternative Program (ACWA); nuclear, chemical and biological treaties; and the cooperative threat reduction program in the countries that are referred to as the former Soviet Union.

With regard to the chemical demilitarization program mission, my primary goals and objectives are: to ensure that the leadership of the U.S. Chemical Demilitarization Program maintains an enhanced culture of safety within the workplace and to confirm our requirements are clearly communicated to, understood by and acted upon by the contractors and not in conflict with the overall chemical demilitarization weapons program; to increase transparency and openness with the public and the international communities and to work for cooperation from special interest groups in accomplishing our mission; and to meet the Chemical Weapons Convention Treaty requirements.

The mission of the U.S. Chemical Demilitarization Program is to destroy all U.S. chemical warfare-related material, while ensuring maximum protection of the public, personnel involved in the destruction effort and the environment. At first glance, this appears to be a fairly straightforward mission, easily achievable provided reasonable resources and effort.

What I have seen since 1986, when Congress mandated the destruction of our chemical weapons stockpile, is there are many unexpected and substantial challenges that must be overcome while conducting a national-scale chemical weapons destruction program. As the GAO recently highlighted, it is obvious that many of these challenges correspond with program management. The Department has already taken—and is in the process of taking—additional steps to rectify program management issues and continue to provide rigorous program oversight.

Recent program changes, today I would like to highlight for you some of the major changes the Department of Defense is implementing with respect to the United States Chemical Demilitarization Program. Earlier this year, the Army consolidated the program manager for chemical demilitarization and its chemical weapons storage mission under the new agency, the U.S. Army Chemical Materials Agency, otherwise known as CMA.

The creation of the CMA allows us to streamline our overall efforts with respect to the chemical weapons destruction and will significantly improve our program management in the long term. This will be accomplished by consolidating accountability, simplifying the chain of command, restructuring and combining organizational functions to reduce redundancies and aligning the program under the checks and balances of the acquisition community through the Assistant Secretary of the Army for Acquisition, Logistics and Technology, with continued oversight by the Office of the Secretary of Defense.

Additionally, the Department of Defense has now selected destruction technologies for all of our chemical weapons sites. On July 16, 2002, the department selected neutralization followed by bio-treatment as the technology to pilot test the destruction of chemical weapons at the Pueblo, Colorado chemical depot.

On February 3, 2003, the department selected neutralization followed by super critical water oxidation as the technology to pilot test the destruction of chemical weapons at the Blue Grass Army depot in Richmond, Kentucky. These chemical weapons destruction sites are currently being managed by the program manager for ACWA as mandated by public law 107–248. The GAO recently found that this division of the program management structure, currently bifurcated between the CMA and the program manager for ACWA, is a significant program deficiency. The Department of Defense agrees with the GAO's assertion.

While the department has operated strictly under the tenets of public law requiring the Office of the Secretary of Defense management of the ACWA program, we are looking at further streamlining the management of the chemical demilitarization program through statutory changes. The program manager for ACWA completed his original charter by successfully demonstrating alternatives to the incineration process. And the department now desires his consolidation under CMA. This consolidation would significantly improve the overall management of the chemical demilitarization program by making the executive agent of the program, the Army, responsible for the program in its entirety. We ask for your full support of this proposal. And we can provide you more detailed information upon your request.

Program status, this year, the chemical demilitarization program has entered a critical phase, with the Aberdeen, Maryland and Anniston, Alabama sites operational. Within the next year, we also expect to commence operations at three additional sites: Umatilla, Oregon; Newport, Indiana; and Pine Bluff, Arkansas. Due to unfortunate circumstances, the Army destroyed little chemical agent over the past year-and-a-half. This is a primary cause of why the

U.S. had to ask the Organization for the Prohibition of Chemical Weapons in September for an extension to the Chemical Weapons Convention intermediate deadline of 45 percent.

I am pleased to report that the organization granted our extension request last week. While we are assured of a high degree of confidence that this will occur by December of 2007, we expect to achieve this milestone sooner. We also expect to have to request an extension for the convention's 100 percent destruction deadline in 2006. The convention allows a maximum five year extension of the 100 percent deadline until April of 2012 at the latest.

As the chemical demilitarization program matures over the next few years, the department will be better prepared to determine the specific length of the extension required. Program issues and concerns, within the next several years, our most significant challenge will likely be the result of our success. As we get our chemical weapons destruction sites online, the Army will have to manage simultaneously up to six separate sites, each operating 24 hours a day, 7 days a week, in four separate time zones, destroying multiple agents, with different technologies and different contractors. As you could have surmised, this will be a significant challenge for the Army, though we are confident they will perform this task exceptionally.

Other factors could also present formidable challenges in our chemical weapons destruction program. Although we take every known precaution to prevent them, accidents and safety incidents may occur, sometimes crippling our destruction efforts. We have considerably strengthened our safety program to mitigate any accidents to counter any future risks. We also face continued opposition from special interest groups through litigation. Requirements for the chemical stockpile emergency preparedness program, or CSEPP, continue to grow.

The Federal Emergency Management Agency and the Army will continue to validate the states' CSEPP requirements and ensure that they meet the maximum protection criteria codified by U.S. Code 50, Chapter 32, Section 1521. And our environmental permitting and monitoring requirements can influence our program. These regulatory changes introduce improvements which affect baseline costs and schedules, though presently this is a manageable key component of our overall chemical weapons destruction program. Finally, at this critical time, in which many destruction facilities are coming online, effective resource management will be a critical influence over our overall destruction strategy.

In conclusion, I want to emphasize the department's intention to address the chemical demilitarization program management issues underscores our commitment to strengthening and improving the overall organizational effectiveness. Changes have already begun at the top, with future changes expected to positively impact all aspects and levels of program management. We have many distinctive challenges; however, we are also poised to work each and every issue to bolster our overall efforts in this prominent national security program.

I welcome your comments on all aspects of the program management. I thank you, Mr. Chairman and the committee, for the opportunity to testify today. I look forward to working with you to advance out common goal of the safe and complete destruction of our nation's chemical weapon stockpile.

[The prepared statement of Mr. Wakefield can be found in the Appendix on page 58.]

Mr. SAXTON. Mr. Wakefield, thank you very much. It is a great statement.

May I just ask, in the interest of time, we will make all of your statements part of the record in their entirety. If you could kind of summarize for us, we would appreciate that.

Mr. Bolton.

STATEMENT OF HON. CLAUDE BOLTON, JR., ASSISTANT SECRETARY OF THE ARMY (ACQUISITION, LOGISTICS AND TECHNOLOGY)

Secretary BOLTON. Mr. Chairman and members of the committee, I am Claude Bolton. And I am grateful to have the opportunity to address this committee.

As the Assistant Secretary of the Army for Acquisition, Logistics and Technology and the Army Acquisition Executive, I am responsible to the Acting Secretary of the Army and to the defense acquisition executive for the execution of the chemical demilitarization program. I consider it an honor to serve in this capacity and to lead the program at this critical juncture when operations have just begun or soon will begin at most of the destruction facilities.

Be assured that I am fully committed to successful implementation of the chemical demilitarization program. I am also committed to ensuring that the public, the Congress, the Secretary of the Army and other senior Department of Defense leadership have timely, accurate information about the progress of the program.

This past February, the Secretary of the Army transferred the secretariat-level oversight and overall program responsibilities from the Assistant Secretary of the Army, Installations and Environment to the Assistant Secretary of the Army, Acquisition, Logistics and Technology, the ASA (ALT).

With that transfer, the former program manager for the Chemical Demilitarization Organization was merged with the former Soldier Biological and Chemical Command Storage and Security Organization into, as was indicated earlier, CMA, the U.S. Army Chemical Materials Agency, headed by Mr. Mike Parker, who you will hear from next.

This organization is jointly overseen by myself and by the commanding general of the Army Materiel Command. Under the new organizational structure, I retain overall program responsibility and will maintain oversight over all phases of the program.

With that reorganization, I am confident that we are moving forward with an organizational structure that enables us to safely and efficiently rid the Nation of these outdated weapons. In my written statement, you will find a diagram, which outlines the organization structure.

As you are aware, in last year's defense authorization act, it mandated that the program continue to be managed as a major defense acquisition program. Be assured that I intend to follow that direction specifically. And in addition to maintaining emphasis on cost, schedule and performance, as well as safety, I will ensure that

Defense Acquisition Workforce Improvement Act has certified personnel to manage the program throughout its life cycle. The mission of the chemical demilitarization program is to destroy all U.S. chemical warfare materiel, while ensuring maximum protection to the public, program personnel and the environment. As pointed out earlier, this is an exciting and successful time for the program.

I am proud to inform you that our first destruction facility on Johnston Atoll in the Pacific successfully destroyed over 2,000 tons of agent, completing its mission in November of 2000. Closure ceremonies for that facility will take place next week. And that will commemorate the end of our mission in the Pacific. In addition, we have three plants in operation, destroying the nation's stockpile of chemical agents and munitions and expect to have three of the remaining five online next year. We have made tremendous strides in community protection in the past few years, ensuring the local communities are fully prepared. Each day that a chemical destruction facility operates, the threat posed to the public by continued storage is reduced.

Since the chemical demilitarization program began, we have safely destroyed over 26 percent of the nation's stockpile, which originally included over 31,000 tons. Presently, incineration facilities for chemical weapons destruction are operating at Tooele, Utah, Anniston, Alabama and a neutralization facility is in operation at Aberdeen, Maryland. Our incineration facilities at Umatilla, Oregon and Pine Bluff, Arkansas are complete and undergoing systemization. These facilities are scheduled to become operational in calendar year 2004. The Pine Bluff, Arkansas facility was completed ahead of schedule and under budget. We have truly learned from the past. Construction is nearly complete at our neutralization-based facility at Newport, Indiana. The two remaining stockpile sites at Pueblo, Colorado and Blue Grass, Kentucky have selected technologies under the assembled chemical weapons alternatives program.

I would like to reiterate that our paramount objective is to reduce the risk to the communities surrounding the chemical storage areas as safely and as quickly as possible. We are also committed to the United States' obligations under the Chemical Weapons Convention and will continue to explore any available means to accelerate the destruction of this nation's stockpile safely and effectively.

In closing, Mr. Chairman, I ask for your continued support of this critical national program. That support will demonstrate our commitment to both the communities surrounding our storage sites and our international partners. Thank you for the opportunity to present my statement to you and the members of this committee. And I look forward to responding to your questions.

[The prepared statement of Secretary Bolton can be found in the Appendix on page 65.]

Mr. SAXTON. Thank you very much, Mr. Bolton.

Mr. Parker, would you proceed?

STATEMENT OF MICHAEL PARKER, DIRECTOR, U.S. ARMY CHEMICAL MATERIALS AGENCY

Mr. PARKER. Thank you, Mr. Chairman.

Mr. Chairman and members of the committee, I am Mike Parker, director of the U.S. Army Chemical Materials Agency. As Mr. Bolton stated, I am responsible for the safe storage and disposal of the U.S. chemical weapons stockpile and non-stockpile materiel, while providing maximum protection to our workers, the public and the environment.

Within this mission set is the chemical stockpile emergency preparedness program that enhances emergency response capability on our storage installations and, in conjunction with FEMA, supports the off-post emergency response community. Additionally, through the cooperative threat reduction program, CMA provides chemical weapons destruction assistance to Russia.

CMA, while a relatively new organization, will resolve some of the issues raised by the General Accounting Office. CMA has already clarified oversight relationships and removed fragmentation hampering execution of the program within the department's capability. As Mr. Wakefield outlined, there are still some provisions in law, which separate the assembled chemical weapons program from the Chemical Materials Agency program. While those separations do cause some inefficiencies in execution, we do come together under Mr. Wynne for the purposes of the overall execution of the program at the Office of Secretary of Defense (OSD) level.

CMA has also, since its formation, invigorated the strategic planning process, developed a risk management plan and established some objective metrics to gauge program performance. Since this program's inception, we have had many successes and overcome many challenges, while continuing to focus on maximum protection to the public, the workers and the environment.

Presently, three disposal facilities are operating: at Deseret Chemical Depot in Utah, Anniston, Alabama and Aberdeen Proving Ground, Maryland. Operations at our Anniston facility, our newest large-scale plant, are particularly encouraging. This plant has exceeded expectations, processing significantly more M55 rockets than both Johnston Island and Tooele, at the same operational point, while demonstrating an exceptional safety record. Aberdeen has been very successful in introducing a totally new technology, an innovative use of commercial, off-post waste treatment facilities.

We are in the process of optimizing the facility at Aberdeen for full rate production. Again, Aberdeen has demonstrated the continuum of outstanding safety record of our DEMIL facilities, while accelerating the overall destruction schedule at Aberdeen by approximately two years.

Of particular note is that all three of these very complex facilities that are online now have successfully met extremely stringent criteria established in the Army surety program, as overseen by the Department of Army Inspector General. The inspector general independently assesses safety, security and accountability necessary to safely process these highly lethal materials. And the facilities have demonstrated the capability to meet this extremely high threshold.

These overall operational thresholds and independent assessments argue well for the safe startup and operation of three additional facilities at Newport, Indiana, Umatilla, Oregon and Pine Bluff, Arkansas, that we plan to bring on within the next year. The two remaining sites—Pueblo, Colorado and Blue Grass, Kentucky—fall under the program manager for Assembled Chemical Weapons Alternative programs, as outlined by Mr. Wakefield. I have a duel hat, as both the director of the Chemical Materials Agency and as the program manager for Assembled Chemical Weapons Alternatives, which does bring an element of unification to the programs, even though they are separated by law.

We take pride in having safely destroyed more than 26 percent of the original U.S. chemical stockpile, as each munition destroyed makes our citizens safer. In the emergency preparedness area, all our storage installations, the on-post portion, have achieved full readiness. And we are in a sustainment phase.

We are working very closely with the Department of Homeland Security, FEMA, state and local emergency management agencies to help our communities around the chemical storage sites to enhance their emergency response capabilities. We are also working very closely with FEMA to address outyear emergency response requirements at the state and local level, to try and bring a more disciplined budget process, so that the requests that we bring forward to Congress will be reflective of the requirements in a better understand and a more fully laid out manner.

Overall, we have not been able to move the destruction process as quickly as we originally envisioned. We have been developing cost and schedule estimates for a new comprehensive acquisition program baseline, which the defense acquisition executive has recently approved in April of this year. This acquisition program baseline places the objective for disposal completion at all sites between 2008 and 2011; the exception being the Blue Grass site, in that the site has recently come under contract and we have yet to be able to fully develop an objective schedule for this site.

Due to the technical complexities of the chemical disposal operation and constantly evolving regulation, new interpretation of existing regulations, we will always face unforeseen challenges that make cost growth and schedule extensions hard to avoid. The Secretary of Defense has challenged us to bring and take all available measures to move the program forward as quickly as possible, to remove the risk to our public. And we are working very hard to meet this challenge. Full congressional support of the President's budget request will be absolutely essential in maintaining the progress.

The last thing I would like to discuss is compliance with the Chemical Weapons Convention. The next major milestone—the 45 percent milestone—is in April 2004, with 100 percent destruction milestone in April 2007. The treaty permits extensions approved by the Conference of State Parties to the Organization for the Prohibition of Chemical Weapons. Our current progress will not meet the 45 percent milestone. And as a result, the U.S. has formally requested an extension of the Chemical Weapons Convention. That extension has been agreed to. And the new date is December 2007. We are very confident in meeting this revised date, while maintaining the safety, our paramount concern, of the workforce, the public and thc environment.

Thank you again for allowing me the opportunity to address this committee today. And I stand ready to answer any of your questions.

Mr. SAXTON. Mr. Parker, thank you. Thank you very much.

We will move now to Mr. Conklin.

STATEMENT OF CRAIG CONKLIN, CHIEF, NUCLEAR AND CHEMICAL HAZARDS BRANCH PREPAREDNESS DIVISION, EMERGENCY PREPAREDNESS AND RESPONSE DIVISION, DEPARTMENT OF HOMELAND SECURITY

Mr. CONKLIN. Mr. Chairman and members of the subcommittee, I am Craig Conklin, and I am chief of FEMA's Nuclear and Chemical Hazards Branch, which is responsible for the CSEPP program at FEMA headquarters.

My comments today will focus on: FEMA's roles and responsibilities in the program; recent improvements in community preparedness; and our commitment to the future of the CSEPP program. FEMA's mission is and has always been to protect our citizens from natural disasters and technological hazards. This mission has not changed with our integration into the Department of Homeland Security.

With respect to the CSEPP program, FEMA's mission is to provide maximum protection to the citizens living in the communities around these facilities. FEMA and the Army have defined "maximum protection" as the avoidance of fatalities to the maximum extent practicable. This means that the efficacy of any protective measure must be considered in light of the benefit and cost associated with both the design and implementation of that protective measure. Under the existing memorandum of understanding with the Army, FEMA is responsible and accountable for all off-site emergency preparedness activities.

First, we validate requests and activities and administer off-post CSEPP funds to support them. We support state and local officials in developing response plans. Third, we develop, deliver and evaluate training. Fourth, we provide technical assistance. And fifth and last, we develop programs for evaluating off-site readiness capability.

In 1993, the Army and FEMA established national program benchmarks, which we use to measure the program performance. These benchmarks serve as a basis for funding decisions, establishment of program priorities and all the primary means by which we manage program performance. FEMA and Army personnel, working closely together with state and county officials, have resolved most of the remaining critical emergency preparedness issues off-site. For example, in June of last year, Oregon Governor Kitzhaber certified that the emergency plans for the communities around the Umatilla Chemical Depot were adequate and fully operational, thus allowing the Army to proceed with test burns.

In Alabama, we successfully addressed the critical preparedness issues of overpressurization of schools, protection of special needs populations and notifications of citizens living near the facility. We must continue to work closely with our state, county, tribal nation partners to ensure that these outstanding issues are addressed and to ensure that a high level of preparedness is maintained throughout the life of the program at each site.

The success that we have experienced in the last year-and-a-half is directly related to the outstanding collaboration and cooperation among FEMA headquarters and regional staff and the Army's new Chemical Materials Agency and its predecessor organization and the off-site communities.

FEMA is fully committed to ensuring that this level of collaboration and cooperation continues. And we will be updating our memorandum of understanding to codify that. The CSEPP program is effectively accomplishing its mission. Nearly all systems are in place and operational. Plans and procedures have been developed and exercised. And public education activities are reaching their intended targets.

Soon, all of these sites will reach the sustainment phase, the phase at which all program benchmarks will have been met. We will need to work just as hard, however, to ensure that these systems stay operational, the plans and procedures stay up to date and the public is kept informed of activities at the sites.

In conclusion, FEMA is firmly committed to ensuring the successful implementation of the CSEPP program and protecting the health and safety of our citizens. The maximum protection standard is the most stringent requirement of any emergency preparedness program or directive. But we are confident that the federal, state and local emergency management community is up to the challenge.

We are extremely thankful to the communities for their commitment and dedication to this important work and pledge to work with them until the risk of a chemical stockpile incident no longer exists. We look forward to that day, when the last chemical weapon and warfare agent is destroyed. Thank you again for allowing me the time to address this committee today. And I look forward to your questions.

[The prepared statement of Mr. Conklin can be found in the Appendix on page 71.]

Mr. SAXTON. Mr. Conklin, thank you very much also. Those are all great statements.

Let me just begin by asking unanimous consent for the statement of Mr. Walden of Oregon and questions that he has for the record be entered in the record at the close of today's hearing.

Mr. Parker, Secretary Bolton and Mr. Wakefield, if you could take a crack at answering this question for us? In the 13 years since the beginning of the pilot plant operations in Johnston Atoll, the Department of Defense and the Army have spent approximately $6 billion to destroy approximately 26 percent of the U.S. chemical weapons stockpile, which originally amounted to over 31,000 tons of chemical agents and munitions.

Current estimates are that 9 years and very close to $20 billion additional will be needed to complete destruction of the stockpile. And the time and cost required to do so could increase as requirements are defined for destruction of the stockpile at Blue Grass, Kentucky, using neutralization technology demonstrated at the assembled chemical weapons assessment and as a result of the factors affecting program costs and schedules cited by GAO and DOD inspector general in their reports.

This is a lot of money. If the number turns out to be $20 billion, that is $2.2 billion a year. And of course, we have lots of uses to which we can put this amount of dollars.

So it prompts me to ask: what measures are the Department of Defense and the Army considering that might reduce the cost and shorten the time required to complete the destruction of the stockpile? And in addition, are there any additional measures which would require action by Congress in order to help you to accomplish these goals of shortening the time and lessening the number of dollars that are required?

Why don't we start with Mr. Parker on this one?

Mr. PARKER. Always like to start with an easy one like this, Mr. Saxton. That is a very challenging question. It is a very challenging program. And the history of the program does speak to the challenges and why the program has experienced the cost and schedule growth that is has. Having said that, we have learned from our experience at Johnston Island and Tooele.

The overall lifecycle cost estimates are adjusted accordingly. We put in place, within those schedule risk and cost risk, recognizing a high level of uncertainty in the program, we put in place risk mitigation measures to address some of the known risk and experience probable risk and develop mitigation programs to try to contain the cost and the schedule growth in the future.

We have taken measures—as demonstrated at Newport, Indiana and Aberdeen, Maryland—to pursue what we call "speedy neutralization," which has rapidly accelerated those two sites, as far as the ability to destroy the agent. While our primary thrust was the safety aspect in accelerating disposal at those two sites, it did offer an opportunity to reduce overall program costs by a couple of hundred million dollars.

We are constantly on the look for these kinds of opportunities. One, I might note, is an initiative that came out of Mr. Wakefield's office to once again look at what we can do from a program standpoint, the programmatic side and the technical side, to reduce cost and schedule. Also, I think out of that same effort, and Mr. Wakefield, I am sure, will address it more, are actions that we might ask the Congress to take that would create opportunities to accelerate the program, to reduce the risk to the public, but also might offer some opportunity for cost savings as well.

Mr. WAKEFIELD. Mr. Chairman, as Mr. Parker indicated, after 9/11, it became exceedingly important for the department not only to look at the safe and secure destruction of the existing stockpile, but also the additional enhanced security. That, indeed, took place.

While at the same time, we looked back at the stockpile sites and determined that the accelerated destruction at the Aberdeen, Maryland site as well as the Newport, Indiana site were not imprudent steps. Indeed, as you know, they have taken place now. We are not satisfied. And it really goes back to the heart of your concern, which is: how do you continue to contain costs or indeed, in fact, reduce costs?

We have conducted in an informal sense what I would refer to as an accelerated destruction, what we call working WIPT, integrated product team, which brings together the full depth and breadth of a number of people, with their skills and talents, to continue to look upon the entire spectrum of the program, to look for efficiencies and effectiveness. We also use a group within the Department of Defense, which is a cost analysis improvement group, to indeed look at both the schedule issues, the cost issues and operational aspects, to look at efficiencies and define what they are. We do this in a cooperative manner, rather than a punitive manner, with the program manger, because that is a key to the success of trying to implement any number of them.

Specific issues, in terms of what we may ask the Congress, we are not ready for, or, if you will, fully prepared at this time, because the work, if you will, is still in progress. But I will assure you that we looked at the entire program. And we looked across everything from just simple process changes to what are laws that we may indeed to seek through the administration, to the Congress, to ask for improvements to help the program in and of itself. But these are still work in progress at this point and not fully characterized at this time.

Mr. SAXTON. Have you identified anything that Congress needs to do to help contain costs?

Mr. WAKEFIELD. Not right now, at this specific time. I think our greatest issue that we have with cost and trying to contain issues seems to be the CSEPP program.

That seems to be, I would refer to, probably the most challenging area. The largest reason is because, of that particular program, the department, as you know, years in advance prepares budgets, reevaluates those budgets and then ultimately submits them as part of the President's submission to the Congress as his budget.

The problem we have is in the year of execution, when requirements come about. And that is where the troublesome issues occur. And we have to juggle the resources across that. Thus far, they have been manageable. But they do intend to be challenging issues.

Mr. SAXTON. Mr. Bolton, what do you think?

Secretary BOLTON. Well, Mr. Chairman, as I had noted in my testimony, I took this program over from another assistant secretary a few months ago. When I first took this position, over a year-and-a-half, almost two years ago, I was approached to take this program over. And I resisted. I looked at the history. I saw a number of things that the GAO has already pointed out in their report. And I resisted until I was assured that were I to take the leadership of this, that we could put some things in place. The chemical demilitarization program needs to be run as a program.

It needs to have a streamlined management. It needs to have a clear focus. It needs to have goals and objectives. And it needs to measure each and every one of those. Once I had those assurances, I took the responsibility. My task has been, since I took that responsibility, to baseline this program: one, to put an organizational structure together that makes sense; next, to put a plan together that says, within that organizational structure that you saw in my testimony, what the heck goes on in each one of those boxes.

Who is doing what? Where is the accountability? Where is the responsibility? What do all those lines mean? And how does that get us toward the goal, which is to safely and quickly get rid of out-

Mr. SAXTON. Thank you very much.

Dr. Snyder.

Dr. SNYDER. Thank you, Mr. Chairman.

I wanted to ask, looking through your opening statements, you know, if we did not have all five of you here and just took selected statements, you would come away with different impressions of where the program is at. And I am going to ask you all to clarify it.

First, Mr. Bolton and Mr. Wakefield; Mr. Bolton, in your opening statement, you say in your written statement, "This is an exciting and successful time for the program. We have three plants in operation, destroying the nation's stockpile of chemical agents and munitions and expect to have three of our five remaining sites operational by the end of next year. Each day a chemical destruction facility operates, the threat posed the public by continued storage is reduced."

That sounds very upbeat. Then I read Mr. Wakefield's statement, which he read. He said, "Due to unfortunate circumstances, the Army destroyed little chemical agent over the past year-and-a-half."

Now, you know, we have a reporter over here who is going to report to the people of Pine Bluff, Arkansas. If they took one of these statements, they would say, "This is great, an exciting and successful time for the program." The other one says, "unfortunate circumstances." And GAO is sitting in the middle, and they have to ask him. I mean, did we have unfortunate circumstances? Or did we have an exciting and successful past 18 months? What is the answer, Mr. Wakefield and Mr. Bolton? Are you in agreement or disagreement?

Secretary BOLTON. I think it is kind of feeling which part of the elephant that you are on. I think it is exciting simply from the standpoint of where we are going in the future.

We have three sites already in operation. Three of the five remaining will be operational over the next year. We are doing it safely. And we are disposing of material. And I will let Mr. Wakefield answer the other. We did have stoppage at one plant. And we had to do that in order to upgrade it.

Dr. SNYDER. So you agree with Mr. Wakefield's comment that——

Secretary BOLTON. Oh, yeah.

Dr. SNYDER [continuing]. Due to unfortunate circumstances, we have destroyed little chemical agent over the last year? So when we talk about it being an exciting and successful time, we are looking ahead to the party, not at what has happened in the last——

Secretary BOLTON. That is absolutely correct.

Dr. SNYDER. All right.

Secretary BOLTON. Yes, sir.

Dr. SNYDER. Also, Mr. Bolton, in your statement about chemical weapons, it says, "We are also committed to meeting the United States' obligation under the Chemical Weapons Convention and will continue to explore available means to accelerate the destruction of the nation's stockpile safely."

I mean, I think a fair reading of that would say, okay, we are on track. And yet, it is obviously clear that we are behind, that we are having to ask for extensions. It is not clear that we are going to make the deadline. I mean, these seem to be again, they seem to be in conflict, if you were just the average reader. And there are a whole lot of people in these communities that are average readers that are trying to figure out what this means.

These statements seem to be in conflict. I do not think I need a response on that one because we have already acknowledged we have had to get extensions. And we will probably have to get other extensions. And it is not clear that we are going to meet the 100 percent deadline. Is that a fair statement, Mr. Wakefield? That it is not clear we will meet the 100 percent deadline?

Mr. WAKEFIELD. In terms of looking at the calendar year 2007, we will not make the 100 percent destruction deadline.

Dr. SNYDER. We will not.

Mr. WAKEFIELD. We have informed the international organization of that. And we will take those steps, as provided by the convention, and properly provide our extension request, based on the solid information we will have as time matures forward.

Dr. SNYDER. Then just a third example, and this may be unfair, but in Mr. Bolton's statement, he uses the construction of the Pine Bluff facility as an example of how we are ahead of schedule. But then, Mr. Hinton, in your statement, you point out that because of diversion of money, that the Pine Bluff program is 6 months behind schedule.

And again, you know, how we report to the community, I mean a fair reading of Mr. Bolton, your statement would send a message to Pine Bluff we are ahead of schedule when, in fact, we are not ahead of schedule. I think they have done a good job down there.

I think it is Congress that messed with them on the diversion of money primarily. But are we in agreement that the Pine Bluff program is behind schedule, but for reasons essentially out of their control? Are you, Mr. Hinton and Mr. Bolton, in agreement with that?

Secretary BOLTON. Once again, it is the elephant that we are dealing with. And it depends on which schedule you are referring to. And I do not mean to be splitting hairs here. But there was an original schedule. And there is the schedule we are looking at now.

My whole intent, when I said earlier that we are going to baseline this, is to put out one schedule, that the GAO and myself and others can look at over time and say: where are we? And so I can report back to you next year, when we come down to the schedule, there is only one schedule we are looking at, not two or three. And therefore, my testimony is based upon one. GAO may have looked at another. And we are all confused.

A long way of saying you are absolutely right. It is conflicting. It is going to be confusing to those who read the testimony. And it is up to people like me to make some changes so it is all consistent.

Dr. SNYDER. Well, thank you, Mr. Chairman. It is gratifying to know that you all did not get together and plan your statements ahead of time. That is clear from these. Thank you.

Mr. SAXTON. Thank you, Dr. Snyder.

Mr. Kline?

Mr. KLINE. Thank you, Mr. Chairman. Thank you, gentlemen, for being with us today.

I share the concern of my colleagues and Americans about how long this process is taking and how much it costs. And in order to keep the costs down and make the progress as quick and as safe as possible, we clearly need an organization to make that work. And I am looking, Secretary Bolton, at your statement. And there is a diagram, figure one, that you have submitted. And in it, it says that the Assistant Secretary of the Army for Acquisition, Logistics and Technology (ASA(ALT)/AAE) reports to the Defense Acquisition Executive (DAE). And the DSAECW reports to the ASA and has oversight of the director of U.S. Army Chemical Materials Agency (CMA), who works simultaneously for the ASA(ALT)/AAE and the CGAMC.

Mr. SAXTON. Thank you for clarifying that. [Laughter.]

Mr. KLINE. It makes me wonder if this is really the most efficient and simplest organization that we could have to get this done. And even without the acronyms, it does appear to be pretty complicated, with coordination and oversight and dual chains or perhaps triple chains.

You expressed some confidence in your statement that this is an organization that will work. And I do not want you to have to refute that claim. But do you think that there is possibly a better way to do this?

And then I would like to ask Mr. Hinton if he has some opinion about this organization as well.

Mr. SAXTON. Before you answer, if I may just announce, we are going to have a series of votes beginning at about 10:30. Those votes will likely last the better part of a half hour. So if we can move through the questions that remain between now and the time we go to vote, it will make all of our days a little easier. So thank you.

Secretary BOLTON. Mr. Kline, first of all, you are absolutely right. Clean sheet of paper, we would write another organization. Given where we were a year ago and given the cultures and the disparate organizations that we were dealing with, this is not a bad first step in trying to consolidate this. I believe in the future there will be some additional changes.

Now the changes that you see there, and having two lines, there is actually a point—we call it a transition to operations—where General Kern, who is the current commanding general for U.S. Army Material Command (AMC), and I will sit down with the team and say, "Okay. We have finished building this plant. We have finished the systemalization of the plant."

"We are now ready to start the operation on these particular chemicals. What is going to happen? How are we going to do that? What people are involved?" And once that goes over, then General Kern, who owns all of these plants, I mean, it is on his books. A lot of the people out there belong to him, on the supporting side.

And so he will have the day-to-day oversight of that, as he does right now. Ultimately, the Secretary wanted one person to point at for all of this. So the Secretary, my boss, does not look at this as, "Well, I have General Kern over there and Bolton over here." He only looks at Bolton. And so it is incumbent upon that dotted line,

General Kern and I, to be in each other's pocket on a regular basis. And we are. We were last night, for example.

I mentioned earlier that we are working a plan. The title here says, Chemical Demilitarization Program "(CDP) Management Plan." This is part of a plan.

The plan says what goes on in each one of those boxes, what all those lines are and eventually what the metrics are to see whether or not this organization really functions. What those metrics allows me to grow to the next iteration of this organization.

But right now, it requires the personalities to work very closely together. I am aiming for an organization that works well, regardless of who is sitting in the boxes.

Mr. KLINE. Thank you.

Secretary BOLTON. Yes, sir.

Mr. KLINE. Mr. Hinton, do you have an opinion here?

Mr. HINTON. Mr. Kline, the ultimate responsibility for the organization is going to rest with the Department. GAO has commented about the complex organization in several reports.

Stability of leadership has been an issue over the years. I think the sooner they can get an organization that is streamlined, gets the leadership in place and has a clear strategy, a clear implementation plan with good metrics of how well we are going to be monitoring our progress against the goals, and most importantly, everybody's roles and responsibilities are laid out. And when you talk about this program, it has to bring all the pieces together, not just what is in the Army, but what is outside of the Army, into a comprehensive plan. That is where we have been in terms of our reporting.

Mr. KLINE. Okay. Thank you both.

I hope that the personalities are working out. But I, like you, would much prefer an organization that would work independent of that. And I hope that we move to something that is, in fact, simpler and more streamlined and I will not have to twist my tongue to figure out who is in charge of who. And I see my time is expired.

Thank you.

Mr. SAXTON. Mike Rogers?

Mr. ROGERS OF ALABAMA. Thank you, Mr. Chairman. Thank you for holding these hearings.

To let you know the frame of reference I have, I live about two miles away from the Anniston facility. The threat that that stockpile poses is a very real threat in my life and the life of my family and my community. So I am keenly interested in this subject.

In listening and reviewing your report, your statement, Mr. Hinton, I noticed that you had made a pretty thorough analysis and had some pretty critical remarks about DOD and the Army. Have you made a similar analysis of the CSEPP program? Because I did not see it referenced in your statement.

Mr. HINTON. Yes, we have, Mr. Rogers. We have issued several reports on the CSEPP program over the years, talking about some of the concerns that we have seen, but basically trying to encourage the Army and FEMA to be more proactive in working together as a team with the community and also in dealing with the requirements of the communities to make sure they are addressed.

We have several reports that we have issued, a lot of recommendations. And both departments have pretty much been in agreement with our recommendations and what we have seen as movement to try to improve in that arena. And I recognize that as these sites get ready to become operational, there are going to be new requirements coming out of the communities. And that is going to have to be part of the proactive role that DOD and FEMA play with the communities to make sure all the needs get addressed well.

Mr. ROGERS OF ALABAMA. I agree. Those analyses, has there been a particular analysis that was specific to the CSEPP program and its role in the Anniston facility?

Mr. HINTON. I believe all of the sites were included in our last report. I will have to check that out. I do not have that with me right now.

Mr. ROGERS OF ALABAMA. Could you have somebody on your staff forward a copy of that report to my office?

Mr. HINTON. I sure will. I sure will. Be happy to.

Mr. ROGERS OF ALABAMA. And Mr. Conklin, my experience has been that the facility in Anniston was fully functional and ready to operate long before it went online a couple, 3 months ago. You made a statement earlier that you felt like, at this time, everything was a go, as far as the CSEPP program was concerned, including the special needs program. Is that your understanding now?

Mr. CONKLIN. Yes, sir. We believe things are in good shape down there. That does not mean that everything is finished. There are things that we still need to follow up on.

Mr. ROGERS OF ALABAMA. For example?

Mr. CONKLIN. For example, on the communications system, we need to make sure that the 800 megahertz system is replaced and operational. These systems have a finite life. And this is going to be true at all of the facilities. So radio systems, alert notification systems, we are going to need to make sure that they maintain their level of operation throughout the life of the program.

Mr. ROGERS OF ALABAMA. So you have identified the special needs population and you feel comfortable that you can reach out to them and handle any problem that might arise?

Mr. CONKLIN. We feel good at how the county set that up, how Calhoun County set up a program to address those, and the funding that went with it and how they implemented that.

Mr. ROGERS OF ALABAMA. Do you know offhand what year you started receiving funding or started providing funding for the CSEPP program in the Anniston community?

Mr. CONKLIN. Not right offhand. I would have to go back and check. I have been with the program since June of last year. So I would have to go back and check the records on that.

Mr. ROGERS OF ALABAMA. Do you know roughly about how much money has been spent on the CSEPP program for the Anniston facility?

Mr. CONKLIN. Yes, sir, I do. I have a table here. For the State of Alabama, we have spent approximately $177 million for that site.

Mr. ROGERS OF ALABAMA. Could you provide me an itemization of how much money has been spent and for what? Because my figure is almost double that, that I understand has been spent.

Mr. CONKLIN. Yeah, okay. I will get that information.

Mr. ROGERS OF ALABAMA. In making preparation for the future, as we go forward with the incineration process, which is now moving forward and at a rate faster than expected, I am pleased to say, how long do you anticipate your program being necessary? And how much more money do you think will be necessary to operate it?

Mr. CONKLIN. Our program will be necessary as long as there are materials stored there or being disposed of.

Mr. ROGERS OF ALABAMA. And that timeframe is what?

Mr. CONKLIN. The latest information I have, that program could go out to another decade or so. I do not know the exact date, but maybe after the year 2016, 2014, somewhere in that vicinity.

Mr. ROGERS OF ALABAMA. And what do you project your budgetary needs to be for that period of time?

Mr. CONKLIN. From fiscal year 2004 to fiscal year 2016, looking at about $145 million, based on what we know now.

Mr. ROGERS OF ALABAMA. Okay. Thank you very much.

Mr. SAXTON. Thank you, Mr. Rogers.

Mr. Cooper? Mr. Larsen?

Mr. LARSEN. Thank you, Mr. Chairman. I walked in after the testimony. But I do not think it was addressed at all by anyone. And I think when I walked in, Mr. Parker was answering a question. You mentioned chemical non-stockpiles. And I am not sure who to speak to.

I do not know who best to answer it, but I would like to get an understanding of where chemical non-stockpile fits into this, the locating, securing and eliminating the non-stockpiles. How does it fit into the mission here? And what timeline exists for these actions? It is part of the CWC? Or is it a separate activity that we have to undertake as well. And I am not sure who in the hierarchy needs to answer that, but I would like to get an answer.

Mr. Parker.

Mr. PARKER. Mr. Larsen, the non-stockpile program covers a number of areas. It covers recovered chemical munitions. Those are munitions that were buried, for whatever reason, and recovered. And those have been declared under the terms of the treaty. Those munitions need to be disposed of by the 2007 timeframe.

Other aspects——

Mr. LARSEN. Are those included then in the 31,000 or so metric tons that were originally identified?

Mr. PARKER. They are declared in the U.S. that 31,000 tons was what we call the stockpile portion. The non-stockpile portion is also declared under the treaty. It is a very, very small number in tonnage, but a significant challenge, nonetheless, because of the condition of the munitions. The non-stockpile program also covers the former production facilities that were used to make chemical weapons. Those need to be destroyed.

The binary weapons which were a late production of a new class of chemical weapons, they are also being destroyed under the non-stockpile program. There is also the facilities which produced feed materials, feed stock chemicals or munition components for chemical weapons, which are also destroyed under the non-stockpile program.

Mr. LARSEN. Does all this testimony that we have heard today apply then only to the stockpile, chemical stockpile program?

Mr. PARKER. Under the total U.S. chemical munitions program, of which the roughly $24 billion encompasses the lifecycle costs, includes the non-stockpile program as well.

Mr. LARSEN. Okay. Does the identification of the chemical non-stockpile include also—you mentioned chemical stockpiles that were buried. But how about those that were dumped in the oceans?

Mr. PARKER. It is——

Mr. LARSEN. Specifically the U.S.

Mr. PARKER. It covers it in the sense that the program——

Mr. LARSEN. We know we did it. We can not get to it?

Mr. PARKER. Well, the program had a responsibility to identify all the locations we could where munitions were buried or where they were dumped. As far as within the $24 billion, to go in and recover those munitions that were dumped in the ocean or the totality of munitions that were buried, that is not within the cost. It is only the recovered portion of the munitions are within that cost.

Mr. LARSEN. Is there an effort to try to locate and recover those dumped?

Mr. PARKER. Yes. In fact, this fiscal year, fiscal year 2004, there has been a significant plus-up in the program to identify and characterize burial sites. There are 200 burial sites that have been identified in 38 states. We are now in the process of trying to characterize what is in those individual sites.

Mr. LARSEN. That is good to know. My question was related to the chemicals dumped in the ocean.

Mr. PARKER. At this point in time, other than identifying where and the quantities, that is the limit.

Mr. LARSEN. Okay. I have a question about the bio-treatment and neutralization and what do you call it, super critical water oxidation at Colorado and Kentucky. My understanding is that the Assembled Chemical Weapons Alternatives Program is responsible for that, the neutralization programs there. And do any of you all represent that, can answer a question about that?

Mr. PARKER. That is one of the two hats that I have. One is the director of the Chemical Materials Agency. The other is the program manager for the ACWA program.

Mr. LARSEN. Okay. Are these totally new technologies that we will be using?

Mr. PARKER. The neutralization part is a well-established technology that has been around for many, many years. The treatment, bio-treatment or super critical water oxidation, is a relatively new technology, although we are using it very successfully for mustard agent at Aberdeen Proving Ground for the bio-treatment.

Mr. LARSEN. And the critical water?

Mr. PARKER. Super critical water oxidation is, in and of itself, it is an established technology. Applying it to chemical weapons waste is a new technology, a totally new application.

Mr. LARSEN. I see my time is up. Just want to make a quick point about that and just to put a marker, if you will, in the record, that we may want to take a look at that to see how things are going, considering this new application of this technology, so that we are not coming back in the future and having this hearing all over again, only having it about the new technologies that we are applying on neutralization.

Mr. PARKER. We will make note of that. And it is covered in the annual report to Congress on the Chem-Demil program. So we do report to the Congress on an annual basis on those kinds of issues.

Mr. LARSEN. Thank you very much. Thank you.

Mr. SAXTON. Thank you, Mr. Larsen.

Ms. Davis?

Ms. DAVIS OF CALIFORNIA. Thank you. Thank you, Mr. Chairman. Thank you all for being here.

I am sorry I missed some of the testimony, but we had worked on a few questions beforehand and I wanted to just address those. One is in process and one on security. And the process question really goes to the incineration plants and the recognition that a majority of the delays were occurring in the incineration plants, identified for various environmental or community reasons. And in the testimony, it did not appear that there were not other types of delays in the neutralization sites.

And so I was wondering why there are so many more procedural delays at the incineration sites and why we are not moving toward a more total neutralization process? Perhaps you have already addressed that or a similar issue?

Mr. PARKER. The history of permitting incinerators was rather contentious. And it was difficult. And it took longer than we had originally estimated to obtain operating permits—construction and operating permits or RCRA permits, Resource Conservation Recovery Act permits for the incinerators.

The neutralization sites, the four neutralization sites were approached in a slightly different manner, with a much higher level of public involvement and public outreach, as a lesson learned from the incinerator plants which had preceded the neutralization plants. And I think a combination of bringing the public in and get a better understanding of what the technologies were, plus possibly a stronger comfort level with the public with neutralization, led to less contentious permitting. Permitting at the neutralization sites has been more straightforward.

As far as the operational aspects, what we are finding, because the first neutralization site has just come online at Aberdeen Proving Ground in April of this year, bringing on a new technology has been just as challenging as bringing on the first incinerator. So the neutralization technology has its own issues, as far as operational aspects, and is just as challenging as the incinerators were in addressing those new technology issues. As an example, Johnston Island, our first incinerator, was very challenging to bring online.

Our newest facility, which we just brought on in August at Anniston, Alabama, an incineration facility, has gone relatively smoothly and has learned from the prior experience at Johnston Island and Tooele, Utah. So I think the issue of one technology being better than the other from an operability standpoint does not bear out in the facts.

Ms. DAVIS OF CALIFORNIA. So there were other technologies that were tried or looked at early on?

Mr. PARKER. Well, the program back in the early 1980's, before the decision was made to proceed with the incineration process, looked at a full spectrum of technologies. Per the direction of the Congress, the National Research Council was the oversight body doing independent assessment to advise the Congress and assist the Army in making a technology look.

So the technology was exhaustively looked at before the incineration was picked in the mid–1980's. Now technology has progressed over the years and allowed us to pursue neutralization followed by bio-treatment or super critical water oxidation for later application. But at the point of time that the decision was made to proceed with incineration, that was a fully vetted process that looked at a wide spectrum of technologies and made the choice of the best option possible at that point in time.

Ms. DAVIS OF CALIFORNIA. Thank you. Just a question on security. I note that in August, a lab tech in Oregon was able to take home a vial of solution containing serum. The question: are there security breaches that are continuing? And how do we address those issues? Perhaps again you have already talked about that.

Mr. PARKER. The specific incidence in Oregon was what is called a laboratory standard, which is a highly dilute material, which is not considered a chemical warfare agent. And it is not even considered a risk. It is basically a dilute alcohol solution. It has a few parts per billion of a reference material in it for calibrating instruments.

The individual, nonetheless, should not have been taking it home. It was a breach of operational procedure, rather than a security breach. Having said that, we have instituted a lesson learned out of that, which involved much stricter adherence and supervisory oversight and accountability of these reference materials, treating them almost in the same manner as we treat the actual chemical warfare agent itself, which has a very, very stringent control procedures to maintain security of those materials.

Mr. SAXTON. Ms. Davis, I wonder, if you have any additional questions or if other members have additional questions, perhaps we could submit them in writing for the record?

Ms. DAVIS OF CALIFORNIA. Thank you.

Mr. SAXTON. And as you can hear by the bells and buzzers, we are going to have to run off and catch these three votes. So rather than have you sit here for a half an hour, 45 minutes to wait for us to get back, we will say, "Thank you."

We will certainly look forward to the progress that you are able to make, both in terms of dollar savings and time shortening. And we look forward to working with you. And we invite you and ask you and hope that when you see something that Congress can do to be helpful, in terms of changing whatever it is that needs to be changed, that you will not hesitate to let us know. Thank you very much. And we appreciate you being here today.

[Whereupon, at 10:40 a.m., the subcommittee was adjourned.]

APPENDIX

OCTOBER 30, 2003

PREPARED STATEMENTS SUBMITTED FOR THE RECORD

OCTOBER 30, 2003

OPENING STATEMENT OF JIM SAXTON
CHAIRMAN, TERRORISM, UNCONVENTIONAL THREATS, AND CAPABILITIES SUBCOMMITTEE
HOUSE ARMED SERVICES COMMITTEE

HEARING ON DESTRUCTION OF THE U.S. CHEMICAL WEAPONS STOCKPILE – PROGRAM STATUS AND ISSUES
OCTOBER 30, 2003

Today, the Terrorism, Unconventional Threats and Capabilities Subcommittee meets to review the Department of Defense program for destruction of the U.S. stockpile of lethal chemical warfare agents and munitions. We are joined in the hearing by several members of Congress who have chemical stockpile storage sites in their districts and who are interested in the chemical demilitarization program: Gentlemen, welcome.

The U.S. chemical weapons stockpile originally consisted of approximately 31,000 tons of lethal chemical agents in a wide variety of munitions, located at Johnston Atoll in the Pacific southwest of Hawaii and eight sites in the continental United States.

The Fiscal Year 1986 Defense Authorization Act requires that the destruction of the stockpile be carried out so as to insure maximum protection of the environment, the general public, and the workers at the storage and demilitarization sites. Destruction of the stockpile began at Johnston Atoll in 1990 and is supposed to be completed by April 29, 2007 in accordance with the Chemical Weapons Convention Treaty.

As of September 24, 2003, approximately 8,220 tons or 26 percent of the stockpile has been destroyed. Our witnesses will talk about progress in the chemical demilitarization program, but will also talk about issues related to program execution, management, cost and schedule.

As members of Congress we will hear good news and bad news, today. The good news will the progress being made in the program as new chemical weapons destruction facilities are brought on line, weapons are destroyed, and safety of the public and the environment increased as the stockpile is reduced. The bad news will be estimates of increased time to complete the destruction program and the increased program costs and implications for US commitments under the Chemical Weapons Convention Treaty that will result.

The total cost of the program has grown from an estimated $1.7 billion in 1986 when the program was initiated at the direction of Congress to an estimated $25 billion today. The estimated time to complete destruction of the stockpile now extends into the next decade. A number of key factors have affect the ability of the Department of Defense and the Army to effectively control the cost and schedule of the chemical stockpile destruction program, many of which are not under the control of either organization. The chemical demilitarization program is very large and complex, and is influenced by a number of offices and entities within and outside the Department of Defense – not the least of which has been Congress.

The issue for Congress and for this subcommittee, today, is to gain an understanding of the progress being made in the chemical weapons destruction program, factors affecting the growth in program schedule and cost, and what might be done to accelerate the completion of the program, both from the standpoint of reducing the cost of the program and ensuring the maximum protection of the public, the personnel involved in destruction of the stockpile, and the environment.

Our witnesses today include

Mr. Henry L. Hinton, Jr. from the General Accounting Office,

Mr. Patrick Wakefield, Deputy Assistant to the Secretary of Defense (Chemical Demilitarization and Counterproliferation),

The Honorable Claude M. Bolton Assistant Secretary of the Army (Acquisition, Logistics, and Technology),

Mr. Michael A. Parker, Director of the U.S. Army Chemical Materials Agency, and

Mr. Craig Conklin from the Federal Emergency Management Agency

Gentlemen, welcome. We look forward to your testimony.

Before our witnesses begin I want to yield to the distinguished ranking member of the subcommittee, Mr. Marty Meehan of Massachusetts for any opening statement he may wish to make.

Mr. Meehan

(Mr. Meehan's Opening Statement)

Thank you, Mr. Meehan.

Mr. Hinton, you'll be our first witness, but before you begin I would ask unanimous consent to enter into the record a recent report by the Department of Defense Office of the Inspector General, "The Chemical Demilitarization Program: Increased Costs for Stockpile and Non-Stockpile Chemical Materiel Disposal Programs", Report No. D-2003-128, dated September 4, 2003. The report discusses the factors that continue to affect the cost and schedule of the Chemical Stockpile Disposal Program and copies of the report are available on the Inspector Generals' web site.

Hearing no objection, so ordered.

Mr. Hinton, you may proceed.

Testimony of Congressman Greg Walden

Subcommittee on Terrorism, Unconventional Threats and Capabilities

October 30, 2003

Chairman Saxton and members of the Terrorism, Unconventional Threats and Capabilities Subcommittee, I appreciate the attention being shown here today to the critical issue of chemical weapons demilitarization. As one who represents the Umatilla Chemical Weapons Depot in eastern Oregon, the process by which the U.S. Army and the federal government, working in concert with states and local communities, works to destroy the nation's chemical munitions is of the highest importance to me. Given the gravity of the threat to the people of Oregon posed by these chemical agents, both I and my colleagues in this Congress have a profound responsibility to ensure that our citizens are properly prepared as we move toward the complete destruction of the nation's chemical weapons stockpile. For this reason, it is paramount that we spare no expense in preparing our communities to respond quickly and effectively to the accidental release of chemical agents.

Mr. Chairman, as you know the Umatilla Chemical Weapons Depot is the next such facility in the United States to begin the process of incinerating its chemical munitions, which is scheduled to commence in early-to-mid 2004. As this date approaches, community leaders in Morrow and Umatilla counties have expressed concern that their community preparedness program stands in jeopardy of not being completed in a timely manner. In a July 23 letter to Oregon Governor Ted Kulongoski, members of the Morrow County Court detailed how this possibility stems from the insufficient provision of funds by the federal government for current and future requirements under the Chemical Stockpile Emergency Preparedness Program (CSEPP).

Gov. Kulongoski forwarded these concerns to Under Secretary Michael D. Brown of the Federal Emergency Management Agency and Under Secretary R.L. Brownlee of the Department of the Army. The Governor's letter stated that, "projected funding in FY 2004 for [Oregon's] CSEPP efforts will likely fall dramatically short of...funding needs." Indeed, the Life Cycle Cost Estimate forecasts $5.6 million for Oregon CSEPP, a figure that is far short of Oregon's FY 04 request for $12.9 million, which itself was lower than the original request of $14.2 million.

The current proposed figure of $5.6 million, if left unchanged, could delay essential emergency preparedness projects even as the final preparations for the commencement of incineration are being made. Among these are:

- continuation of an evacuation project for the City of Hermiston
- evacuation buses for local school children
- upgrades to existing over-pressurized schools
- additional warning sirens and tone alert radios for the Immediate Response Zone

- enhanced respiratory protection equipment and decontamination supplies for first responders, and
- a maintenance contract for a new 450 MHz tactical communications radio system.

Far from representing extravagant requests, these projects are fundamentally necessary to the safety of the communities situated near the Umatilla Chemical Weapons Depot. It is my view that any failure to implement these essential projects would be an unconscionable abdication of our responsibilities as public officials. Moreover, given the extraordinary lengths to which the federal government went to accommodate the State of Alabama prior to the commencement of chemical agent incineration, I am hopeful that the government will recognize its duty to provide an equitable level of protection for all citizens and thus treat the State of Oregon similarly.

In response to complaints from Morrow County officials that this sum was insufficient, Mr. Craig Conklin, Chief of the Nuclear and Chemical Hazards Branch Preparedness Division at the Department of Homeland Security, responded that estimates of local CSEPP funding needs were "generated by…our state and county partners, and then validated by FEMA and the Department of Defense (DoD) for inclusion in the President's budget." This suggestion was hotly disputed by the local officials I represent. Indeed, in an October 15, 2003, response to Conklin they wrote, "Research indicates that neither the State of Oregon nor local government in the Umatilla Community EVER approved the Army's initial Life Cycle Cost Estimate figure." Their letter goes on to state that the LCCE "has never sought to identify Umatilla Community needs, but enforces an inflexible and unrealistic decade-old 'overly optimistic estimate.'"

As the members of the Morrow County Court made clear in a prior letter to Mr. Conklin on August 18, 2003, concerns have also arisen over a statement in the CSEPP Cooperative Agreement (CA) guidance of July 14, 2003 that calls into question the prudence of proceeding with the incineration of weapons at the Umatilla Depot as scheduled. The CA guidance states, "Use of the four CSEPP National Quantitative Performance Indicators is not required in FY 04." This statement seems to contradict the CSEPP National Benchmark, as listed in FEMA's CSEPP Policy Paper No. 18, which lists four of these benchmarks that are "consider[ed] fundamental to successfully preparing for the necessary public protection." These four benchmarks are: 1) capability to provide timely warning to the population at risk; 2) knowledge of the appropriate response by the public to warning system sounds; 3) coordination of emergency plans between and among the Army installation and response agencies; and 4) ensured functioning and reliability of communications systems. Given the incomplete nature of both the evacuation project for the City of Hermiston and the 450 MHz radio communications system, the question must be asked whether the federal government considers the Umatilla community to be adequately prepared for the commencement of agent incineration.

Like the overwhelming majority of the citizens of Morrow and Umatilla counties, I favor the swift destruction of chemical weapons at the Depot as soon as we are able to proceed safely and effectively. However, adequate public safeguards must not be sacrificed for the sake of efficiency in destroying these munitions. As members of the Morrow County Court have stated, "While we do strongly support incineration to remove this known risk to our community, we don't do so at the expense of safety and protection to our first responders and citizens." It is of the utmost importance that the concerns of both Morrow County officials and Governor Kulongoski be addressed by federal officials, lest disagreements over funding and preparedness jeopardize the planned commencement of incineration in 2004.

Finally, community leaders in Morrow and Umatilla County have stressed the importance of having officials in Washington, D.C. visit the communities where chemical weapons facilities are based. In particular, they are hopeful that Mr. Craig Conklin will accept their invitation to visit the Umatilla Depot prior to the commencement of agent incineration to hear firsthand from the officials and first responders charged with safeguarding the surrounding communities.

Thank you again for giving me the opportunity to register my concerns with the Subcommittee.

United States General Accounting Office

GAO

Testimony

Before the Subcommittee on Terrorism, Unconventional Threats and Capabilities, Committee on Armed Services, House of Representatives

For Release on Delivery
Expected at 9 a.m. EST
Thursday, October 30, 2003

CHEMICAL WEAPONS

Better Management Tools Needed to Guide DOD's Stockpile Destruction Program

Statement of Henry L. Hinton, Director,
Defense Capabilities and Management

GAO
Accountability * Integrity * Reliability

GAO-04-221T

October 30, 2003

GAO
Accountability·Integrity·Reliability

Highlights

Highlights of GAO-04-221T, a testimony to the Subcommittee on Terrorism, Unconventional Threats and Capabilities, Committee on Armed Services, House of Representatives

CHEMICAL WEAPONS

Better Strategic and Risk Management Tools Needed to Guide DOD's Stockpile Destruction Program

Why GAO Did This Study

Since its inception in 1985, the Chemical Demilitarization (Chem-Demil) Program has been charged with destroying the nation's large chemical weapons stockpile. After years of planning and building new facilities, the program started destroying the stockpile in 1990. As of October 2003, the program had destroyed 26 percent of the 31,500-ton agent stockpile, and its total estimated cost to destroy the entire stockpile is more than $25 billion.

This testimony summarizes GAO's September 2003 report and addresses the following issues:
(1) the status of schedule milestones and cost estimates,
(2) the impact of the current schedule on the Chemical Weapons Convention (CWC) deadlines,
(3) the challenges associated with managing the program, and (4) the status of the Chemical Stockpile Emergency Preparedness Program (CSEPP).

What GAO Recommends

GAO recommended in its September 2003 report that the Department of Defense (DOD) and the Army develop an overall strategy and implementation plan for the program and implement a risk management approach, and DOD concurred.

www.gao.gov/cgi-bin/getrpt?GAO-04-221T.

To view the full product, including the scope and methodology, click on the link above. For more information, contact Henry L. Hinton, Jr. at (202) 512-4300 or hintonh@gao.gov.

What GAO Found

The Chem-Demil Program faces schedule delays and higher costs, but it has improved emergency preparedness in communities near the sites. In 2001, the Chem-Demil Program extended its schedule milestones and increased its cost estimates from $15 billion to about $24 billion. Since then nearly all sites have experienced delays, stemming from problems such as: plant safety issues, environmental requirements, approving emergency preparedness plans, and funding shortfalls. The program needs a risk management plan to mitigate problems affecting program schedules, costs, and safety. Program officials say the delays have raised the cost estimates by an additional $1.4 billion, to more than $25 billion as of September 2003. Based on current schedule slippages, GAO believes that costs will grow higher and further delays will occur. (See figure.)

Because of schedule delays, the United States will not meet CWC's April 2004 deadline to destroy 45 percent of the stockpile and it risks not meeting the original 2007 deadline to complete destruction of the entire stockpile. Unless the program fixes the problems causing delays, the United States also risks not meeting CWC's deadline of 2012, if extended.

The program has suffered from several long-standing management and organizational issues. The lack of sustained leadership has undercut decision-making authority and obscured accountability. The program's complex structure, with multiple lines of authority, has left roles and responsibilities unclear. It does not have an overarching, comprehensive strategy to guide and integrate its activities and monitor its performance.

The Army and the Federal Emergency Management Agency have helped state and local communities become better prepared to respond to chemical emergencies. Despite these gains, CSEPP costs are rising because some states have expanded their preparedness requests beyond the approved budgets. These requests amount to $88 million for fiscal years 2004 and 2005.

Comparison of 1998, 2001, and 2003 Cumulative Program Cost Estimates

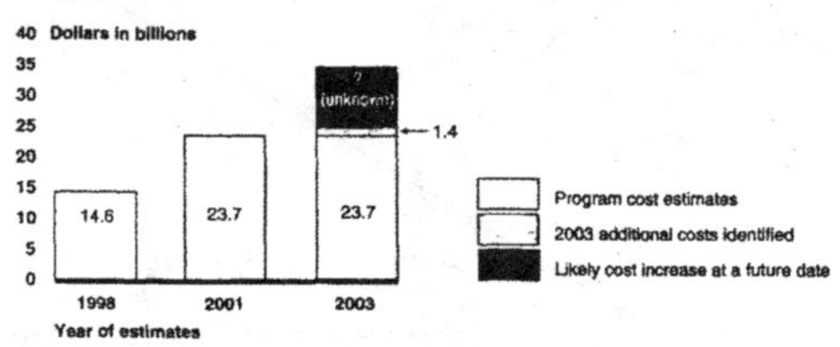

Source: DOD.

Mr. Chairman and Members of the Subcommittee:

I appreciate the opportunity to participate in this hearing today on the Department of Defense's (DOD) Chemical Demilitarization (Chem-Demil) Program. Since its inception in 1985, this program has been charged with destroying the nation's large chemical weapons stockpile, second only to Russia's in terms of its size. After years of planning and building new facilities, the program started destroying the stockpile in 1990.

As you requested, my statement focuses on the following issues: (1) the status of schedule milestones and costs at the sites, (2) the impact that the current schedule may have on the Chemical Weapons Convention[1] (CWC) deadlines, (3) the challenges associated with managing the program, and (4) an update on the status of the Chemical Stockpile Emergency Preparedness Program (CSEPP).

As of October 2003, the Chem-Demil Program had destroyed an estimated 8,210 tons (26 percent) of the total 31,500 tons of the original agent stockpile stored at nine sites in the United States and the Pacific Ocean at Johnston Atoll. Of the four sites that have begun agent destruction operations, Johnston Atoll has destroyed all of its stockpile; Tooele, Utah, has reduced its stockpile by about 44 percent; Anniston, Alabama, has destroyed about 2 percent of its stockpile; and Aberdeen, Maryland, has eliminated over 3 percent of its stockpile. Current schedule estimates show that the Army will not complete destruction of the entire stockpile until after the year 2012.

Since 1990, we have issued more than 25 reports on the Chem-Demil Program. Nearly half of the reviews have raised questions about the program's growing costs, its inability to meet its schedule milestones, and its management weaknesses.

[1] In April 1997, the United States Senate ratified the Convention on the Prohibition of the Development, Production, Stockpiling and Use of Chemical Weapons and on Their Destruction, commonly known as the Chemical Weapons Convention. S. Res. 75, April 24, 1997.

My testimony today draws heavily from our most recent report, which was issued in September 2003.[2] In summary, we found the following:

- While the Chem-Demil Program has revised its schedule milestones and increased its cost estimates several times, with the latest revision in 2001, the program still cannot meet them. According to current Army schedules for destruction, five sites will miss their 2001 schedule milestones, less than 2 years after they were reset. The other four sites have not yet missed schedule milestones, but they too have experienced delays. Most of the substantial delays have stemmed from problems that DOD and the Army have been unable to anticipate or influence. These include plant safety issues, difficulties in meeting environmental permitting requirements, public concerns about emergency preparedness plans, and funding shortfalls. Neither DOD nor the Army has adopted a comprehensive risk management approach that could help mitigate potential problems that affect program schedules, costs, and safety by anticipating problems and developing mitigation plans. Army officials told us they are now developing such an approach. According to program officials, the delays that have occurred since the 2001 schedule revisions, along with the resolution of emergency preparedness issues, have raised the program's cost estimates by $1.4 billion, to a current total of more than $25 billion. We expect this amount will grow substantially before the destruction of the stockpile is complete if these delays continue.

- Because of schedule delays, the United States will not meet CWC's April 2004 deadline to destroy 45 percent of the chemical stockpile. The United States recently asked the governing body of the convention for an extension beyond the April 2004 deadline. If the delays that the program has experienced continue, the United States also risks not meeting the 2007 deadline to destroy 100 percent of the stockpile. Unless the Chem-Demil Program is able to fix the problems that have caused these delays, the United States also risks not meeting CWC's deadline, if extended to 2012, to destroy the entire stockpile. The CWC allows extensions of up to 5 years to the 2007 deadline.

- Despite recent efforts to improve the management and streamline the organization of the Chem-Demil Program, the program has suffered from several long-standing and unresolved leadership, organizational, and strategic planning issues. The program has lacked sustained leadership at

[2] U.S. General Accounting Office, *Chemical Weapons: Sustained Leadership, Along with Key Strategic Management Tools, Is Needed to Guide DOD's Destruction Program,* GAO-03-1031 (Washington, D.C.: Sept. 5, 2003).

both the upper levels of oversight and at the program-manager level, which undercuts decision-making authority and obscures accountability.[3] In addition, the program's complex management structure, with multiple lines of authority within the Army and the separation of program components among the Army, DOD, and the Federal Emergency Management Agency (FEMA), has left roles and responsibilities for the different parts of the program unclear. FEMA manages the emergency preparedness program (CSEPP) for communities near the storage sites. Finally, the absence of an overarching, comprehensive strategy has resulted in a program without a clear road map to closely guide and integrate all of its activities and monitor its performance.

- Since our 2001 report,[4] the Army and FEMA have helped state and local communities become better prepared to respond to chemical emergencies. Based on the states' self-assessments and FEMA's reviews, all of the states with nearby chemical storage sites are considered close to being fully prepared for emergency issues. However, despite these accomplishments, CSEPP costs continue to rise because some state and local communities have expanded their emergency preparedness requests beyond their approved budgets, exceeding them by $88 million for fiscal years 2004 and 2005, especially as they move closer to agent operations phase. FEMA and the Army have implemented a number of recommendations we made to improve technical assistance and guidance, training, and compliance measures to assess preparedness.

Most Sites Will Miss Schedule Milestones due to Program's Inability to Anticipate and Influence Issues

Despite several revisions to schedule milestones since the program's inception, the Chem-Demil Program still is unable to meet these milestones because of unanticipated delays. Most incineration sites have missed important milestones established in 2001. Delays at Anniston, Umatilla, and Pine Bluff have already resulted in their missing the 2001 schedule milestones to begin chemical agent destruction operations (operations phase).[5] Johnston Atoll has missed its schedule milestone for shutting down the facility (closure phase).[6] Although Tooele has not

[3] Upper level refers to the offices of the assistant secretary or above in the Departments of the Army and Defense.

[4] *Chemical Weapons: FEMA and Army Must Be Proactive in Preparing States for Emergencies,* GAO-01-850 (Washington, D.C.: Aug. 13, 2001).

[5] At the time of the 2001 schedule revision, all three of these sites were in the systemization phase; thus, their next milestone was to begin agent destruction operations.

[6] At the time of the 2001 schedule revision, agent destruction operations had been completed and its next milestone was to complete closure of the facility.

missed any milestones since the 2001 schedule was issued, the site has undergone substantial delays in destroying its stockpile primarily because of a safety-related incident in July 2002. If additional delays occur at the Tooele site, it could also exceed its next milestone as well. Table 1 shows the status of the incineration sites that will miss 2001 schedule milestones.

Table 1: Slippage of 2001 Scheduled Milestone Dates, by Incineration Site

Site	Next schedule milestone	2001 schedule date to begin next milestone	Date to begin next phase[a]	Difference between 2001 schedule and estimate (no. of months)
Anniston	Operations	July 2002	Aug. 2003	+13
Umatilla	Operations	July 2003	Mar. 2004	+8
Pine Bluff	Operations	Oct. 2003	Apr. 2004	+6
Johnston Atoll	End of closure	Sept. 2003	Nov. 2003	+2

Sources: DOD and the U.S. Army.

[a]Program manager's official estimates for Pine Bluff, Umatilla and Johnston Atoll.

Many of the recent delays at the incineration sites have resulted from operations incidents, from environmental permitting problems, community protection concerns, and funding issues—a trend that we identified in previous reports on the program. Among the events that have caused delays at incineration sites since 2001 are the following:

- *Incidents during operations.* At Tooele, a chemical incident involving a plant worker who came into contact with a nerve agent while performing routine maintenance led to the suspension of agent destruction operations from July 2002 to March 2003. An investigation attributed the incident to inadequate or poorly followed worker safety procedures, and a corrective action plan, including an improved safety plan, was instituted before operations resumed. Since operations restarted in March 2003, Tooele has experienced several temporary shutdowns.

- *Environmental permitting.* Several environmental permitting issues have delayed the start of agent destruction operations at sites at Umatilla and Anniston.[7] At Umatilla, the delays stemmed from several unanticipated

[7] We have reported on permitting delays in *Chemical Weapons And Materiel: Key Factors Affecting Disposal Costs and Schedule,* GAO/NSIAD-97-18 (Washington, D.C.: Feb. 10, 1997).

engineering changes related to reprogramming software and design changes that required permit modifications and to a shutdown by state regulators because furnaces were producing an unanticipated high amount of heavy metals during surrogate agent testing. At Anniston, delays occurred because state environmental regulators did not accept test results for one of the furnaces because the subcontractor did not follow state permit-specified protocols.

- *Community protection.* Concerns about emergency preparedness for local communities have led to additional delays at Anniston. These concerns included the inadequacy of protection plans for area schools and for special needs residents (e.g., elderly and disabled individuals) who would have difficulty in an evacuation. Although we reported on this issue in July 1996[8] and again in August 2001, and a senior DOD official identified it as a key concern in September 2001, the Army had difficulty satisfactorily resolving the issue with key state stakeholders. As a result, operations did not begin until August 2003.
- *Funding.* Delays at Pine Bluff and Johnston Atoll occurred because DOD redirected fiscal year 2002 destruction program funds to acquire $40.5 million worth of additional emergency protection equipment for Anniston. To cover this unfunded budget expense, the Army reduced Pine Bluff's budget by $14.9 million and Johnston Atoll's budget by $25.1 million, leading to systemization and closure milestone slippages, respectively, at these sites. Program officials told us that the total cost of this schedule slip would ultimately be $116 million due to the extended period before closure. The program is likely to face unfunded requirements as programwide funding requests continue to exceed budgeted amounts. As of October 2003, according to preliminary estimates from FEMA, unfunded CSEPP requirements for all sites are expected to amount to $39.4 million and $49.0 million for fiscal years 2004 and 2005, respectively.

Unlike the incineration sites, the two bulk-agent only sites, Aberdeen and Newport, have experienced delays but have not breeched their schedule milestones. In 2002, DOD approved using an alternative technology (neutralization), instead of incineration, at these two sites. This technology is expected to accelerate the rate of destruction at these two sites. The

[8] See U.S. General Accounting Office, *Chemical Weapons Stockpile: Emergency Preparedness in Alabama Is Hampered by Management Weaknesses,* GAO/NSIAD-96-150 (Washington, D.C: July 23, 1996).

Army estimated that this process would reduce the scheduled end of operations at both sites by 5 years, from 2008 to 2003 at Aberdeen and from 2009 to 2004 at Newport. However, Aberdeen has encountered unanticipated problems with the removal of residual agent from bulk containers and has extended its planned completion date by 6 months, from October 2003 to March 2004. In addition, Newport has faced construction delays and community resistance to offsite treatment of waste byproducts. As a result of these delays, Newport has extended its planned start date for agent operations by 5 months, from October 2003 to February 2004.

At two sites, Pueblo, Colorado, and Blue Grass, Kentucky, no milestones were set in the 2001 schedule because DOD had not yet selected a destruction technology. DOD has now selected a destruction technology for these sites, but it made decisions several months later than estimated. More importantly, DOD has set initial schedule milestones for these two sites that go beyond the extended April 2012 CWC deadline. According to DOD officials, these milestones are preliminary and will be reevaluated once contractors finish initial facility designs.

The Chem-Demil Program has faced continued delays with the program largely because DOD and the Army have not yet developed a risk management approach to proactively anticipate and address potential problems that could adversely affect program schedules, costs, and safety. Such an approach could also leverage knowledge of potential problems gained at other sites. Instead, according to a DOD official, the program has used a crisis management approach, which has forced it to react to, rather than control, issues. The program had drafted a plan in June 2000 that was intended to address these issues. However, according to a program official, this plan was never approved or implemented because of a change in management in 2001.

The delays and schedule extensions[9] have contributed directly to program cost growth, according to program officials. As a result, DOD's total program cost estimate grew from $15 billion to $24 billion between 1998 and 2001. (See fig. 1.) Because of delays encountered since the 2001 revisions, the Army is now in the process of developing new milestones that will extend beyond those adopted in 2001. According to an Army

[9] Schedule extensions are caused largely by actual destruction rates being lower than planned.

official, the program will use events that have occurred since 2001 in presenting new cost estimates to DOD for preparation of the fiscal year 2005 budget submission. Program officials told us that they estimate new costs had increased by $1.4 billion as of October 2003, and this estimate is likely to rise further as additional factors are considered.

Figure 1: Comparison of 1998, 2001, and 2003 Cumulative Program Cost Estimates

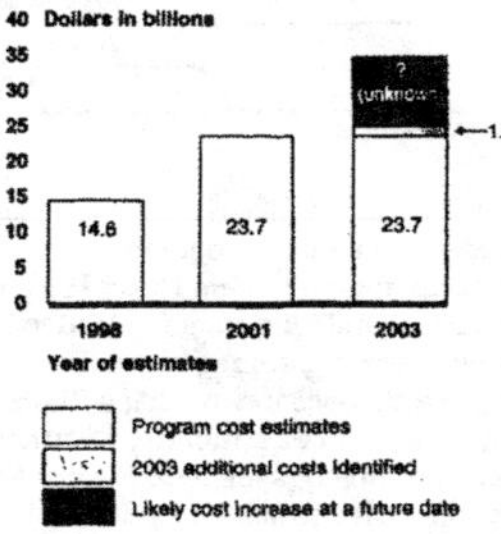

Source: DOD.

Schedule Delays Jeopardize Ability of Program to Meet CWC Deadlines

Although the United States met the first two chemical weapons treaty deadlines, the continuing delays jeopardize its ability to meet the final two deadlines. (See table 2.) Since reaching the 2002 deadline to destroy 20 percent of the stockpile in July 2001, the Chem-Demil Program has been able to destroy only an additional 3 percent of the stockpile. In order to meet the April 2004 CWC deadline to destroy 45 percent of the stockpile, the program would have to eliminate an additional 22 percent of the stockpile within the next 6 months. Because the program will likely not be able to achieve this rate of destruction, the United States has asked for an extension of the 2004 deadline.

According to current destruction schedules, the United States will not meet the 2007 deadline to eliminate 100 percent of the stockpile. As a result, the United States will likely have to ask for an extension of the 2007 deadline to complete the destruction of the entire stockpile. The CWC allows extensions of up to 5 years beyond the 2007 deadline. Unless the program fixes the problems that are causing schedule delays, the United States also risks not meeting this deadline, if extended to 2012.

Table 2: CWC Deadlines

Required percentage of agent destroyed	Deadlines for destruction	Date United States met deadline
1	April 29, 2000	September 1997
20	April 29, 2002	July 2001
45	April 29, 2004	Will not meet
100	April 29, 2007	Will not meet

Sources: CWC and U.S. Army.

Long-standing Management and Organizational Weaknesses Hamper Program Progress

Despite recent efforts to improve the management and streamline the organization of the Chem-Demil Program, the program continues to falter because several long-standing leadership, organizational, and strategic planning weaknesses remain unresolved. The lack of sustained leadership has undercut decision-making authority and obscured accountability. The program's complex structure, with many lines of authority, has left roles and responsibilities unclear. Finally, the program lacks an overarching, comprehensive strategy to guide and integrate its activities and monitor performance.

Leadership Shifts Affect Continuity in Decision Making

The Chem-Demil Program's lack of sustained leadership above the program level is underscored by the multiple shifts in oversight responsibilities that have occurred three times between DOD and the Army during the past two decades. The most recent change took place in 2001 when oversight responsibility for the program shifted back to DOD's Office of the Secretary of Defense. Table 3 summarizes the changes.

Table 3: Transfer of Program Oversight Responsibilities between DOD and the Army, 1986-Present

Year	Oversight authority	Action
1986	Army	DOD designates the Army as the executive agent for the Chem-Demil Program.
1994	DOD	DOD makes the program a major defense acquisition program and oversight is elevated to control cost and schedule increases and to raise program visibility.
1998	Army	DOD delegates decision-making authority to the Army, primarily as part of its overall effort to reduce responsibilities and staffing of its offices.
2001	DOD	DOD reinstates its position as the program's top decision maker. According to DOD, this was done to streamline decision making, which is consistent with the cost of the program and national and state interest in the safe and timely destruction of the stockpile.

Source: GAO analysis of DOD data.

These shifts in oversight responsibilities affected the continuity of program decision making and obscured accountability. As a different office assumed major decision authority, the program's emphasis shifted and initiatives that had been started were often not completed. For example, when the Army had oversight responsibility for the program, it established a memorandum of understanding with FEMA to clarify each of their roles and responsibilities related to CSEPP.[10] However, after DOD assumed the program's oversight responsibilities in 2001, DOD did not follow the protocols for coordination that had been established in the memorandum, according to FEMA and DOD officials. As a result, DOD provided funds for emergency preparedness items without having adequate plans for distribution, which delayed the process. This shift in oversight responsibilities from the Army to DOD also left state and local community officials and other stakeholders uncertain as to the credibility of federal officials. According to FEMA and Army officials, coordination between the two agencies has improved in the last few months and efforts are being made to repair relationships with community and state stakeholders.

[10] U.S. General Accounting Office, *Chemical Weapons: FEMA and Army Must Be Proactive in Preparing States for Emergencies,* GAO-01-850 (Washington, D.C.: Aug. 13, 2001).

Similar problems have also occurred within the Army as program leadership has changed. Three different officials at the Assistant Secretary level have held senior leadership positions since December 2001. In addition, five officials have served as the Deputy Assistant Secretary of the Army (Chem-Demil) during that time.[11] From April 2002 to February 2003, the program manager's position remained vacant for nearly 1 year, before being filled. However, after only 4 months, the program manager resigned and the Army named a replacement.

Frequent shifts in key leadership positions have led to several instances where the lack of continuity affected decision making and obscured accountability. For example, in June 2002, a program official promised to support future funding requests for emergency preparedness equipment from one community, but his successor did not fulfill this promise. Other communities viewed the agreement with one community as an opportunity to substantially expand their own funding requests. The lack of sustained leadership makes it unclear who is accountable when program commitments are made and not fulfilled. Moreover, when key leaders do not remain in their positions to develop the needed long-term perspective on program issues and effectively implement program initiatives, it is difficult to maintain program progress and ensure accountability for leadership actions.

Program Management Structure Remains Complex

As our 2003 report documents, the Army recently reorganized the program. But this change in management structure has not streamlined the program's complex organization nor clarified roles and responsibilities. The establishment of the Chemical Materials Agency (CMA) in January 2003 has left the Director reporting to two different senior Army organizations, which is one more than under the previous structure. This divided reporting approach is still not fully developed, but has the potential to adversely affect program coordination and accountability. The reorganization has also divided the responsibility for various program phases between two offices within CMA. One organization, the Program Manager for the Elimination of Chemical Weapons, will manage the first three phases (design, construction, and systemization) for each site, and a newly created organization, the Director of Operations, will manage the final two phases (operations and

[11] This position is now the Deputy Assistant Secretary of the Army (Elimination of Chemical Weapons).

closure). This reorganization changes the cradle-to-grave management approach that was used to manage sites in the past and has blurred responsibilities for officials who previously provided support in areas such as quality assurance and safety. Moreover, the reorganization did not address two program components—Assembled Chemical Weapons Alternatives (ACWA) program and community-related CSEPP. DOD will continue to manage ACWA separately from the Army, as congressionally directed. In addition, the Army will continue to manage CSEPP jointly with FEMA.

Program Lacks Strategy and Implementation Plan

While DOD and the Army have issued numerous policies and guidance documents for the Chem-Demil Program, they have not developed an overarching, comprehensive strategy or an implementation plan to guide the program and monitor its progress. This is contrary to the principals that leading organizations embrace to effectively implement and manage programs. Some key aspects of an approach typically used to effectively manage programs include promulgating a comprehensive strategy that includes a clearly stated mission, long-term goals, and methods to accomplish these goals. An implementation plan that includes annual performance goals, measurable performance indicators, and evaluation and corrective action plans is also important. According to DOD and Army officials, the Chem-Demil Program has relied primarily on guidance and planning documents related to the acquisition process.[12] However, in response to our recent recommendation that they prepare such a strategy and plan, DOD stated that it is in the initial stages of doing so and estimates completion in fiscal year 2004.

Emergency Preparedness Program Is Improving, but Costs Are Rising

Since our 2001 report, the Army and FEMA have assisted state and local communities to become better prepared to respond to chemical emergencies. Based on the states' self-assessments and FEMA's reviews, all 10 states with chemical storage sites located within them or nearby are now considered close to being fully prepared to respond to a chemical emergency. This is a marked improvement from the status we reported in 2001[13] when 3 states reported that they were far from being prepared. Now, 6 of the 10 states are reporting that their status is fully prepared and the

[12] Acquisition programs establish program goals for cost, schedule, and performance parameters over the program's life cycle.

[13] GAO-01-850.

remaining 4 are close to being fully prepared. However, these statuses are subject to change because the states and communities themselves can revise or expand their agreed-upon emergency preparedness needs. They can make these changes because the "maximum protection" concept that governs CSEPP is open to interpretation. As a result, they can appear to be less prepared than before. For example, Oregon certified that it was fully prepared, but now has requested additional emergency equipment. This request has changed Oregon's self-reported preparedness status from fully prepared to incomplete.

Despite these accomplishments, CSEPP costs continue to rise because, according to Army and FEMA officials, state and local communities may add to their emergency requirements beyond approved requests. Army and FEMA officials explain that the states often identify and expand their requirements, especially as destruction facilities move closer to the start of the operations phase. For example, the states of Colorado, Alabama, and Oregon have all requested funds for infrastructure, including roads and bridges. In June 2002, Oregon certified that its community readiness was adequate and recommended permit approval to allow test burns at Umatilla. Since that time, Oregon has asked for additional emergency preparedness support that exceeds its CSEPP budget. This request follows a pattern of substantially increasing funding requests at the start of the operations phase, as occurred at Anniston in 2001 when it received $40.5 million for additional CSEPP items. Programwide, new requirements continue to exceed approved CSEPP funding levels. FEMA has little control over the additional funding requests made by the states. As of October 2003, FEMA had identified $39.4 and $49.0 million in unfunded requirements for fiscal years 2004 and 2005, respectively. (See table 4.)

Table 4: CSEPP Unfunded Requirements (UFR) for Fiscal Years 2004 and 2005, as of October 2003

Dollars in millions

	FY2004 unfunded requirements			FY2005 unfunded requirements		
Funding entity	**FY2004 funded**	**FY2004 requirements**	**FY2004 UFRs**	**FY2005 funded**	**FY2005 requirements**	**FY2005 UFRs**
Alabama	21.0	41.9	21.0	19.7	36.7	17.0
Arkansas	16.9	24.7	7.8	3.6	13.1	9.6
Colorado	3.1	2.8	-0.3	2.5	4.5	1.9
Confederated Tribes	0.2	3.7	3.5	0.2	0.9	0.7
Illinois	0.9	0.8	-0.1	0.8	0.7	0.0
Indiana	3.9	4.7	0.8	2.3	6.0	3.7
Kentucky	4.3	5.3	1.0	3.7	4.4	0.7
Maryland	2.3	1.5	-0.8	1.8	2.0	0.2
Oregon	5.7	12.4	6.7	4.5	13.9	9.4
Utah	5.6	5.6	0.0	7.1	7.0	-0.1
Washington	5.9	5.8	-0.1	3.1	5.3	2.2
FEMA support	18.7	18.7	0.0	22.6	26.5	3.9
Total	**88.5**	**127.9**	**39.4**	**72.0**	**120.9**	**49.0**

Source: FEMA data.

In our August 2001 report, we recommended that the Army and FEMA (1) provide technical assistance, guidance, and leadership to the three states (Alabama, Indiana, and Kentucky) with long-standing emergency preparedness issues to resolve their concerns; (2) provide all states and their communities with training and assistance in preparing budget and life-cycle cost estimates and provide guidance and plans on reentry; and (3) establish specific measures of compliance with the benchmarks to more evenly assess performance and to correctly identify requirements. The Army is continuing to provide assistance to CSEPP states and communities as requested by FEMA. FEMA now participates more often in local community CSEPP activities and sponsors an annual CSEPP conference in an effort to improve its working relationships. FEMA has also provided software to simplify development of CSEPP financial reporting documents and has published a Reentry and Recovery Workbook. The workbook fills a void in state and local guidance for emergency responders to follow in the event of a chemical emergency. Lastly, FEMA expanded its capability assessment readiness tool to assist local communities in quantifying benchmark scores.

Agencies' Actions to Implement Prior GAO Recommendations

We recommended in our September 2003 report that the Secretary of Defense direct the Under Secretary of Defense for Acquisition, Technology and Logistics, in conjunction with the Secretary of the Army, to (1) develop an overall strategy and implementation plan for the chemical demilitarization program and (2) implement a risk management approach that anticipates and influences internal and external factors that could adversely impact program performance. DOD concurred with our recommendations. It said that it was in the initial stages of developing an overall strategy and implementation plan and estimated that it would be completed in fiscal year 2004. It also said that CMA will review the progress of an evaluation of several components of its risk management approach within 120 days and then that DOD would evaluate the results and determine any appropriate action. In our 2001 report, we recommended that the Army and FEMA make improvements to the program, and they have implemented those recommendations.

Mr. Chairman, this concludes my statement. I would be pleased to respond to any questions that you or members of the Subcommittee may have.

Contacts and Acknowledgments

For future questions regarding this testimony, please contact me at (202) 512-4300. Individuals making key contributions to this testimony include Donald Snyder, Rodell Anderson, Bonita Oden, John Buehler, Nancy Benco, and Mike Zola.

GAO's Mission

The General Accounting Office, the audit, evaluation and investigative arm of Congress, exists to support Congress in meeting its constitutional responsibilities and to help improve the performance and accountability of the federal government for the American people. GAO examines the use of public funds; evaluates federal programs and policies; and provides analyses, recommendations, and other assistance to help Congress make informed oversight, policy, and funding decisions. GAO's commitment to good government is reflected in its core values of accountability, integrity, and reliability.

Obtaining Copies of GAO Reports and Testimony

The fastest and easiest way to obtain copies of GAO documents at no cost is through the Internet. GAO's Web site (www.gao.gov) contains abstracts and full-text files of current reports and testimony and an expanding archive of older products. The Web site features a search engine to help you locate documents using key words and phrases. You can print these documents in their entirety, including charts and other graphics.

Each day, GAO issues a list of newly released reports, testimony, and correspondence. GAO posts this list, known as "Today's Reports," on its Web site daily. The list contains links to the full-text document files. To have GAO e-mail this list to you every afternoon, go to www.gao.gov and select "Subscribe to e-mail alerts" under the "Order GAO Products" heading.

Order by Mail or Phone

The first copy of each printed report is free. Additional copies are $2 each. A check or money order should be made out to the Superintendent of Documents. GAO also accepts VISA and Mastercard. Orders for 100 or more copies mailed to a single address are discounted 25 percent. Orders should be sent to:

U.S. General Accounting Office
441 G Street NW, Room LM
Washington, D.C. 20548

To order by Phone:	Voice:	(202) 512-6000
	TDD:	(202) 512-2537
	Fax:	(202) 512-6061

To Report Fraud, Waste, and Abuse in Federal Programs

Contact:

Web site: www.gao.gov/fraudnet/fraudnet.htm
E-mail: fraudnet@gao.gov
Automated answering system: (800) 424-5454 or (202) 512-7470

Public Affairs

Jeff Nelligan, Managing Director, NelliganJ@gao.gov (202) 512-4800
U.S. General Accounting Office, 441 G Street NW, Room 7149
Washington, D.C. 20548

PRINTED ON RECYCLED PAPER

MAXIMUM PROTECTION OF THE PUBLIC, PERSONNEL INVOLVED IN THE DESTRUCTION EFFORT, AND THE ENVIRONMENT. AT FIRST GLANCE, THIS APPEARS TO BE A FAIRLY STRAIGHTFORWARD MISSION, EASILY ACHIEVABLE PROVIDED REASONABLE RESOURCES AND EFFORT. WHAT WE HAVE SEEN SINCE 1986, WHEN CONGRESS MANDATED THE DESTRUCTION OF OUR CHEMICAL WEAPONS STOCKPILE, IS THERE ARE MANY UNEXPECTED AND SUBSTANTIAL CHALLENGES THAT MUST BE OVERCOME WHILE CONDUCTING A NATIONAL-SCALE CHEMICAL WEAPONS DESTRUCTION PROGRAM. AS THE GAO RECENTLY HIGHLIGHTED, IT IS OBVIOUS THAT MANY OF THESE CHALLENGES CORRESPOND TO PROGRAM MANAGEMENT. THE DEPARTMENT HAS ALREADY TAKEN AND IS IN THE PROCESS OF TAKING ADDITIONAL STEPS TO RECTIFY PROGRAM MANAGEMENT ISSUES AND CONTINUE TO PROVIDE RIGOROUS PROGRAM OVERSIGHT.

RECENT PROGRAM CHANGES

TODAY I WOULD LIKE TO HIGHLIGHT FOR YOU THE MAJOR CHANGES THE DEPARTMENT OF DEFENSE IS IMPLEMENTING WITH RESPECT TO THE UNITED STATES CHEMICAL DEMILITARIZATION PROGRAM. EARLIER THIS YEAR, THE ARMY CONSOLIDATED THE PROGRAM MANAGER FOR CHEMICAL DEMILITARIZATION AND ITS CHEMICAL WEAPONS STORAGE MISSION UNDER A NEW AGENCY, THE U.S. ARMY CHEMICAL MATERIALS AGENCY, OTHERWISE KNOWN AS THE CMA. THE CREATION OF THE CMA ALLOWS US TO STREAMLINE OUR OVERALL EFFORTS WITH RESPECT TO CHEMICAL WEAPONS DESTRUCTION, AND WILL SIGNIFICANTLY IMPROVE OUR PROGRAM MANAGEMENT IN THE LONG TERM. THIS WILL BE ACCOMPLISHED BY CONSOLIDATING ACCOUNTABILITY, SIMPLIFYING THE CHAIN OF COMMAND, RESTRUCTURING AND COMBINING ORGANIZATIONAL FUNCTIONS TO REDUCE REDUNDANCIES, AND ALIGNING THE PROGRAM UNDER THE CHECKS AND BALANCES OF THE ACQUISITION COMMUNITY THROUGH THE ASSISTANT SECRETARY OF THE ARMY FOR ACQUISITION, LOGISTICS AND TECHNOLOGY, WITH CONTINUED OVERSIGHT BY THE OFFICE OF THE SECRETARY OF DEFENSE.

ADDITIONALLY, THE DEPARTMENT OF DEFENSE HAS NOW SELECTED DESTRUCTION TECHNOLOGIES AT ALL OF OUR CHEMICAL WEAPONS SITES. ON JULY 16, 2002, THE DEPARTMENT SELECTED "NEUTRALIZATION FOLLOWED BY BIOTREATMENT" AS THE TECHNOLOGY TO PILOT TEST THE DESTRUCTION OF CHEMICAL WEAPONS AT THE PUEBLO, COLORADO CHEMICAL DEPOT. ON FEBRUARY 3, 2003, THE DEPARTMENT SELECTED "NEUTRALIZATION FOLLOWED BY SUPERCRITICAL WATER OXIDATION" AS THE TECHNOLOGY TO PILOT TEST THE DESTRUCTION OF CHEMICAL WEAPONS AT THE BLUE GRASS ARMY DEPOT IN RICHMOND, KENTUCKY. THESE CHEMICAL WEAPONS DESTRUCTION SITES ARE CURRENTLY BEING MANAGED BY THE PROGRAM MANAGER FOR ACWA AS MANDATED BY PUBLIC LAW 107-248.

THE GAO RECENTLY FOUND THAT THE DIVISION OF THE PROGRAM MANAGEMENT STRUCTURE, CURRENTLY BIFURCATED BETWEEN THE CMA AND THE PROGRAM MANAGER FOR ACWA, IS A SIGNIFICANT PROGRAM DEFICIENCY. THE DEPARTMENT OF DEFENSE AGREES WITH THE GAO'S ASSERTION. WHILE THE DEPARTMENT HAS OPERATED STRICTLY UNDER THE TENETS OF PUBLIC LAW REQUIRING OFFICE OF THE SECRETARY OF DEFENSE MANAGEMENT OF THE ACWA PROGRAM, WE ARE LOOKING AT FURTHER STREAMLINING THE MANAGEMENT OF THE CHEMICAL DEMILITARIZATION PROGRAM THROUGH STATUTORY CHANGE. THE PROGRAM MANAGER FOR ACWA COMPLETED HIS ORIGINAL CHARTER BY SUCCESSFULLY DEMONSTRATING ALTERNATIVES TO THE INCINERATION PROCESS, AND THE DEPARTMENT NOW DESIRES HIS CONSOLIDATION UNDER THE CMA. THIS CONSOLIDATION WOULD SIGNIFICANTLY IMPROVE THE OVERALL MANAGEMENT OF THE CHEMICAL DEMILITARIZATION PROGRAM BY MAKING THE EXECUTIVE AGENT OF THE PROGRAM, THE ARMY, RESPONSIBLE FOR THE PROGRAM IN ITS ENTIRETY. WE ASK FOR YOUR FULL SUPPORT OF THIS PROPOSAL AND CAN PROVIDE YOU MORE DETAILED INFORMATION UPON REQUEST.

PROGRAM STATUS

THIS YEAR THE CHEMICAL DEMILITARIZATION PROGRAM HAS ENTERED A CRITICAL PHASE, WITH THE ABERDEEN, MARYLAND AND ANNISTON, ALABAMA SITES OPERATIONAL. WITHIN THE NEXT YEAR, WE ALSO EXPECT TO COMMENCE OPERATIONS AT THREE ADDITIONAL SITES: UMATILLA, OREGON; NEWPORT, INDIANA; AND PINE BLUFF, ARKANSAS.

DUE TO UNFORTUNATE CIRCUMSTANCES, THE ARMY DESTROYED LITTLE CHEMICAL AGENT OVER THE PAST YEAR AND A HALF. THIS IS A PRIMARY CAUSE OF WHY THE U.S. ASKED THE ORGANIZATION FOR THE PROHIBITION OF CHEMICAL WEAPONS IN SEPTEMBER FOR AN EXTENSION TO THE CHEMICAL WEAPONS CONVENTION INTERMEDIATE 45% DESTRUCTION DEADLINE. I AM PLEASED TO REPORT THAT THE ORGANIZATION GRANTED OUR EXTENSION REQUEST LATE LAST WEEK. WHILE WE ARE ASSURED WITH A HIGH DEGREE OF CONFIDENCE THIS WILL OCCUR BY DECEMBER 2007, WE EXPECT TO ACHIEVE THIS MILESTONE SOONER. WE ALSO EXPECT TO HAVE TO REQUEST AN EXTENSION TO THE CONVENTION 100% DESTRUCTION DEADLINE IN 2006. THE CONVENTION ALLOWS A MAXIMUM FIVE YEAR EXTENSION OF THE 100% DEADLINE (UNTIL APRIL 2012 AT THE LATEST). AS THE CHEMICAL DEMILITARIZATION PROGRAM MATURES OVER THE NEXT FEW YEARS, THE DEPARTMENT WILL BE BETTER PREPARED TO DETERMINE THE SPECIFIC LENGTH OF EXTENSION REQUIRED.

PROGRAM ISSUES AND CONCERNS

WITHIN THE NEXT SEVERAL YEARS, OUR MOST SIGNIFICANT CHALLENGE WILL LIKELY BE A RESULT OF OUR SUCCESS. AS WE GET OUR CHEMICAL WEAPONS DESTRUCTION SITES ON-LINE, THE ARMY WILL HAVE TO MANAGE SIMULTANEOUSLY, UP TO SIX SEPARATE SITES, EACH OPERATING 24 HOURS A DAY, SEVEN DAYS A WEEK, IN FOUR SEPARATE TIME ZONES, DESTROYING MULTIPLE AGENTS, WITH DIFFERING TECHNOLOGIES, AND DIFFERENT CONTRACTORS. AS YOU COULD SURMISE, THIS WILL BE A SIGNIFICANT

CHALLENGE FOR THE ARMY, THOUGH WE ARE CONFIDENT THEY WILL PERFORM THIS TASK EXCEPTIONALLY.

OTHER FACTORS COULD ALSO PRESENT FORMIDABLE CHALLENGES IN OUR CHEMICAL WEAPONS DESTRUCTION PROGRAM:

- ALTHOUGH WE TAKE EVERY KNOWN PRECAUTION TO PREVENT THEM, ACCIDENTS AND SAFETY INCIDENTS MAY OCCUR, SOMETIMES CRIPPLING OUR DESTRUCTION EFFORTS. WE HAVE CONSIDERABLY STRENGTHENED OUR SAFETY PROGRAM TO MITIGATE ANY ACCIDENTS TO COUNTER ANY FUTURE RISKS.
- WE ALSO FACE CONTINUED OPPOSITION FROM SPECIAL INTEREST GROUPS THROUGH LITIGATION.
- REQUIREMENTS FOR THE CHEMICAL STOCKPILE EMERGENCY PREPAREDNESS PROGRAM (OR CSEPP), CONTINUE TO GROW. THE FEDERAL EMERGENCY MANAGEMENT AGENCY AND THE ARMY WILL CONTINUE TO VALIDATE THE STATES' CSEPP REQUIREMENTS AND ENSURE THEY MEET THE 'MAXIMUM PROTECTION' CRITERIA CODIFIED IN U.S. CODE 50, CHAPTER 32, SECTION 1521.
- AND, NEW ENVIRONMENTAL PERMITTING AND MONITORING REQUIREMENTS CAN INFLUENCE OUR PROGRAM. THESE REGULATORY CHANGES INTRODUCE IMPROVEMENTS WHICH AFFECT BASELINE COSTS AND SCHEDULES, THOUGH PRESENTLY THIS A MANAGEABLE KEY COMPONENT OF OUR OVERALL CHEMICAL WEAPONS DESTRUCTION PROGRAM.

FINALLY, AT THIS CRUCIAL TIME IN WHICH MANY DESTRUCTION FACILITIES ARE COMING ON-LINE, EFFECTIVE RESOURCE MANAGEMENT WILL BE A CRITICAL INFLUENCE OVER OUR OVERALL DESTRUCTION STRATEGY.

FINAL REMARKS

IN CONCLUSION, I WANT TO EMPHASIZE THE DEPARTMENT'S INTENTION TO ADDRESS CHEMICAL DEMILITARIZATION PROGRAM MANAGEMENT ISSUES

UNDERSCORES OUR COMMITMENT TO STRENGTHENING AND IMPROVING OVERALL ORGANIZATIONAL EFFECTIVENESS. CHANGE HAS ALREADY BEGUN AT THE TOP, WITH FUTURE CHANGES EXPECTED TO POSITIVELY IMPACT DIFFERENT ASPECTS AND LEVELS OF PROGRAM MANAGEMENT. WE HAVE MANY DISTINCTIVE CHALLENGES; HOWEVER WE ARE ALSO POISED TO WORK EACH AND EVERY ISSUE TO BOLSTER OUR OVERALL EFFORTS IN THIS PROMINENT NATIONAL SECURITY PROGRAM. I WELCOME YOUR COMMENTS ON ALL ASPECTS OF OUR PROGRAM'S PROGRESS. I THANK YOU, MR. CHAIRMAN, AND THIS COMMITTEE FOR THE OPPORTUNITY TO TESTIFY TODAY. I LOOK FORWARD TO WORKING WITH YOU TO ADVANCE OUR COMMON GOAL OF THE SAFE AND COMPLETE DESTRUCTION OF OUR NATIONAL CHEMICAL WEAPON STOCKPILE.

RECORD VERSION

STATEMENT BY THE UNITED STATES ARMY:

THE HONORABLE CLAUDE M. BOLTON, JR

ASSISTANT SECRETARY OF THE ARMY
ACQUISITION, LOGISTICS, AND TECHNOLOGY

BEFORE THE

COMMITTEE ON ARMED SERVICES

SUBCOMMITTEE ON TERRORISM, UNCONVENTIONAL THREATS

AND CAPABILITIES

HOUSE OF REPRESENTATIVES

FIRST SESSION 108TH CONGRESS

CHEMICAL DEMILITARIZATION

30 OCTOBER 2003

NOT FOR PUBLICATION
UNTIL RELEASED BY THE
COMMITTEE ON ARMED SERVICES

STATEMENT BY

HONORABLE CLAUDE M. BOLTON, JR.

ASSISTANT SECRETARY OF THE ARMY

ACQUISITION, LOGISTICS AND TECHNOLOGY

ON THE UNITED STATES CHEMICAL DEMILITARIZATION PROGRAM

Mr. Chairman and members of the Committee, I am Claude Bolton and am grateful to have the opportunity to address this committee. As the Assistant Secretary of the Army for Acquisition, Logistics and Technology and Army Acquisition Executive, I am responsible to the Acting Secretary of the Army and to the Defense Acquisition Executive for the execution of the Chemical Demilitarization Program. I consider it an honor to serve in this capacity and to lead the program at this critical juncture when operations have just begun or soon will begin at most of the destruction facilities.

Be assured that I am fully committed to successful implementation of the Chemical Demilitarization Program. I am also committed to ensuring that the public, the Congress, the Secretary of the Army, and other senior Department of Defense leadership have timely, accurate information about the progress of the Program.

This past February, the Secretary of the Army transferred the secretariat-level oversight and overall program responsibility from the Assistant Secretary of the Army, Installations and Environment (ASA(I&E)), to the Assistant Secretary of the Army, Acquisition, Logistics and Technology (ASA(ALT)). With that transfer, the former Program Manager for Chemical Demilitarization organization was merged with the former Soldier Biological and Chemical Command storage and security organization into a new organization, the U.S. Army Chemical Materials Agency, headed by Mr. Michael Parker. This organization is jointly overseen by the ASA(ALT) and the Commanding General, Army Materiel Command, General Paul Kern. Under the new organizational structure, I retain overall program responsibility and will maintain oversight over all phases of the program. With this reorganization, I am confident that we are moving forward with an organizational structure that enables us to safely and efficiently rid the nation of these outdated weapons.

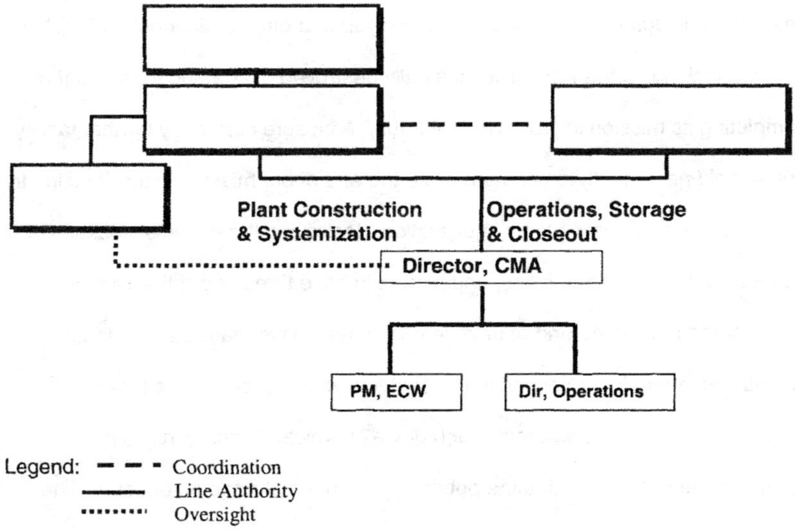

Figure 1. CDP Management Plan

As you are aware, last year's Defense authorization act mandated that the program continue to be managed as a major defense acquisition program. Be assured that I intend to strictly adhere to this direction. In addition to maintaining emphasis on cost, schedule and performance, as well as safety, I will ensure that Defense Acquisition Workforce Improvement Act (DAWIA)-certified acquisition professionals participate fully in the management of the program throughout its life cycle.

The mission of the Chemical Demilitarization Program is to destroy all U.S. chemical warfare materiel while ensuring maximum protection to the public, Program personnel, and the environment. This is an exciting and successful time for the Program. I am proud to inform you that our first destruction facility on Johnston Island, in the Pacific successfully destroyed over 2,000 tons of agent, completing its mission in November of 2000. A closure ceremony for that facility will be held next month to commemorate the end of our mission in the Pacific. In addition, we have three plants in operation, destroying the nation's stockpile of chemical agents and munitions, and expect to have three of our five remaining sites operational by the end of next year. We also have made tremendous strides in community protection in the past few years, ensuring the local communities are fully prepared. Each day a chemical destruction facility operates, the threat posed to the public by continued storage is reduced. The danger of continued storage of these agents and munitions is the impetus that drives the Program forward; the safety of American citizens is the paramount concern for the Army, and indeed, all in attendance today.

Since the U.S. Chemical Demilitarization Program began, we have safely destroyed over 26 percent of the nation's stockpile of lethal chemical agents and munitions, which originally included over 31,000 tons of chemical agent. We are proud of the progress to date, but the true challenges are just beginning. With the commencement of destruction operations across the United States, we will continue to rely heavily on the professionalism of our talented workforce to to perform their duties in a safe and expeditious manner.

Presently, incineration facilities for chemical weapons destruction are operating at Tooele, Utah and Anniston, Alabama and a neutralization facility is operational at Aberdeen, Maryland. Our incineration facilities at Umatilla, Oregon and Pine Bluff, Arkansas are complete and undergoing systemization. These facilities are scheduled to become operational in calendar year 2004. The Pine Bluff, Arkansas facility was completed ahead of schedule and under budget; we have truly learned from the past to better our approaches. Construction is nearly complete at our neutralization-based facility located at Newport, Indiana. The two remaining stockpile sites at Pueblo, Colorado and Blue Grass, Kentucky have selected technologies under the Assembled Chemical Weapons Alternatives Program and will use neutralization followed by bio-treatment and Super Critical Water Oxidation, respectively.

I would like to reiterate that our paramount objective is to reduce the risk to the communities surrounding chemical storage sites as we safely eliminate the United States stockpile of lethal chemical agents and munitions. We are also committed to meeting the United States' obligations under the Chemical

Weapons Convention and will continue to explore any available means to accelerate the destruction of the nation's stockpile safely and effectively.

In closing, Mr. Chairman, I ask for your continued support of this critical national program. That support will demonstrate our commitment to both the communities surrounding our storage sites and our international partners. Thank you for the opportunity to present my statement to you and to members of your Committee. I look forward to responding to your questions.

STATEMENT FOR THE RECORD
CRAIG CONKLIN
CHIEF
NUCLEAR AND CHEMICAL HAZARDS BRANCH
(FEMA)
DEPARTMENT OF HOMELAND SECURITY

Mr. Chairman and members of the Subcommittee, I am Craig Conklin, Chief of FEMA's Nuclear and Chemical Hazards Branch. I am pleased to provide this statement to the Subcommittee wherein I will address FEMA's activities in support of the Chemical Stockpile Emergency Preparedness Program (CSEPP).

We welcome the opportunity to share with the committee many of the CSEPP's successes and how this important program is benefiting our nations' emergency preparedness and homeland security efforts. Following my brief remarks, I will be happy to answer your questions.

My statement will cover: FEMA's roles and responsibilities in CSEPP; the structure and operation of the program; the current status and challenges presented by this complex program; and the efforts to share the lessons learned from this program within the Department of Homeland Security.

FEMA's MISSION

FEMA's mission is, and has always been, to reduce the loss of life and damage to property and to protect our residents from all hazards, natural and man made. We accomplish this mission by providing the Nation with comprehensive, risk-based emergency management programs, including mitigation, preparedness, response, and recovery. This mission has not changed with our integration into the new Department of Homeland Security.

FEMA has many partners that help accomplish this mission. We work closely with many other federal Departments and agencies, with States, Tribal Nations, local governments, volunteer organizations and with private industry.

CHEMICAL STOCKPILE EMERGENCY PREPAREDNESS PROGRAM (CSEPP)

In 1985, Title 14, Part B, Section 1412 of Public Law (P.L.) 99-145, Congress directed the Department of Defense (DOD) to dispose of its lethal unitary (pre-mixed) chemical agents and munitions while providing "maximum protection for the environment, the general public and the personnel involved in the destruction of lethal chemical agents and munitions . . ." While the likelihood of a chemical stockpile incident with off-post consequences is considered remote, the Army and FEMA recognize that the consequences of such an event could be significant.

In response to congressional direction to protect the public during the destruction of the chemical agents, the Army proposed a site-specific emergency planning program for the off-post communities in the ten stockpile states. Thus, the Chemical Stockpile Emergency Preparedness Program or CSEPP was established. FEMA joined the Army in implementing the CSEPP through a Memorandum of Understanding (MOU) signed in August 1988.

Congress mandated "maximum protection" for the environment, general public, and workers in the 1985 law. We should note that the maximum protection standard is the most stringent of any emergency preparedness program.

FEMA and the Army have defined "maximum protection" in CSEPP Policy Paper #1, dated May 7, 1991, as "the avoidance of fatalities to the maximum extent practicable". CSEPP Policy Paper #1 also states "The efficacy of any protective action strategy or preventive measures must be considered in light of . . . the benefit/cost associated with the design and implementation of the protective measure."

CSEPP's mission is to protect residents from exposure to chemical agents and is therefore, directly applicable to the work of DHS and our nation's anti-terrorism efforts. Many of CSEPP's chemical hazard and public safety research, training and public outreach efforts can be used nationwide.

FEMA's CSEPP RESPONSIBILITIES

Management of CSEPP is a unique partnership among the U.S. Army, FEMA, states, Tribal Nations, and local jurisdictions. The Federal management structure is uniquely designed to capitalize on each federal partner's expertise and administrative infrastructure to develop and enhance the emergency preparedness capabilities of the affected Army installations and the participating State and local jurisdictions.

The August 1988 Memorandum of Understanding established the framework for working with affected States, Tribal Nations and local governments to provide for the public's health and safety. FEMA and the Army reaffirmed this original Memorandum of Understanding in January 1993. As the program matured, FEMA and the Army realized that the respective roles and responsibilities needed clarification. Thus, in 1997 FEMA and the Army signed a revised MOU.

Under the terms of this MOU, FEMA has the responsibility and accountability for all aspects of off-post emergency preparedness. Specifically, FEMA:

- Administers off-post CSEPP funds;
- Supports the States in developing response plans;
- Prepares, develops, delivers and evaluates training;
- Provides technical assistance; and
- Develops programs for evaluating off-site readiness capability.

PROGRAM STRUCTURE

Ten States, forty-one counties, and one Tribal Nation surrounding the eight U.S. Army stockpile sites participate in CSEPP. The eight States hosting installations with chemical stockpiles are Alabama, Arkansas, Colorado, Indiana, Kentucky, Maryland, Oregon, and Utah. Illinois and Washington also participate in the program because their borders' proximity to the Indiana and Oregon stockpiles, respectively. The Confederated Tribes of the Umatilla Reservation in Oregon also actively participate in the program.

Thirteen (13) counties are in Immediate Response Zones, the areas closest to where the chemical agents are stored, generally within approximately a ten-mile radius. Twenty-five counties are in Protective Action Zones, beginning at the outer edge of the Immediate Response Zones and extending to a distance of between six and 31 miles. The remaining three counties are designated as host counties, which lie outside the Immediate Response Zones and Protective Action Zones.

The CSEPP is administered through the States, as are FEMA's other emergency preparedness programs. Funds are distributed to the States under Cooperative Agreements, based upon a negotiated work plan between the States and FEMA Regional Offices. Under the agreements, each State identifies needs, develops proposed projects to meet those needs, requests funds, and disburses those funds at the State level and to local governments.

Budgeting for the CSEPP is done according to the Department of Defense's Planning Programming and Budgeting System (PPBS). The budget for off-post emergency preparedness is based in large part on Life Cycle Cost Estimates (LCCE) prepared by the States in conjunction with FEMA and updated regularly.

At the Federal level, FEMA and Army Headquarters are responsible for CSEPP policy and program development, while the FEMA Regions and the U.S. Army's Chemical Materials Agency manage day-to-day operations. Site-specific issues are dealt with through site-specific Integrated Process Teams. These teams (required to be established by Section 1076 of P.L. 104-201, the Department of Defense Authorization Act for 1997) serve as the primary local forum for identifying site-specific operational issues, proposing solutions to those issues to the appropriate level decision makers, and implementing programmatic and operational decisions.

CSEPP MANAGEMENT SYSTEM

CSEPP has been developed from two perspectives. First, it focuses on providing the personnel, equipment, and training necessary to establish a response infrastructure that enables emergency managers to quickly alert the public, manage the response, and communicate with the public, the media, and emergency responders. The second equally important focus is on the public surrounding the stockpile locations: making them aware of what to do in the event of an accident. The cornerstone of CSEPP is the concept of "functional equivalency". Rather than using a formulaic approach (e.g. per capita population) to apportion funds, CSEPP is based on programmatic benchmarks that define a level of response functionality necessary to protect the public or benchmark compliance and provides resources as needed to eliminate preparedness weaknesses.

Fiduciary requirements dictate that FEMA carefully evaluate requests from the States and communities to achieve "maximum protection" capability within the limits of funds provided. As such, our goal is to deliver maximum available resources to the local communities in relation to the level of risk faced by the community.

As of the end of fiscal year 2003, approximately $577 million has been allocated to the States under the annual Cooperative Agreements. In addition, $78.2 million more has been applied to FEMA-managed contracts that support the States. The allocation of resources is tracked according to the jurisdiction that spends the funds rather than the jurisdiction that benefits from the service. Therefore, just looking at the proportion of the funds spent at the State and county levels does not necessarily reflect the true picture of the benefits the counties have received through the program.

CSEPP BENCHMARKS AND PREPAREDNESS COMPONENTS

In May 1993, the Army and FEMA jointly established National CSEPP Benchmarks. The benchmarks identified capabilities to be achieved by participating State and local jurisdictions in meeting the "maximum protection" mandate. These benchmarks also serve as a basis for funding decisions, establishment of program priorities and assessment of capabilities. The benchmarks were updated in August 2000 to add three additional benchmarks and published as CSEPP Policy Paper #18. The current CSEPP National Benchmarks are:

1. Alert and Notification (A&N)
2. Administration (ADM)
3. Automated Data Processing (ADP)
4. Communications (COM)
5. Coordinated Plans (COP)
6. Emergency Operations Centers (EOC)
7. Exercises (EX)
8. Medical (MED)
9. Personnel (PER)
10. Protective Action Strategy (PAS)
11. Public Outreach and Education (POE)
12. Training (TNG)

The Benchmarks are the primary system by which we manage performance in CSEPP. The eight CSEPP communities evaluate and update their benchmark status quarterly at IPT meetings. The FEMA Regional CSEPP personnel then report benchmark status at quarterly In Progress Reviews (IPRs) held at FEMA Headquarters. The CSEPP has made significant strides in improving Benchmark compliance during 2003.

SUCCESSES

The partnership between the U.S. Army and FEMA is currently very strong. Our two organizations have worked well together resulting in numerous accomplishments. Building on those successes, we are working on other initiatives that are designed to enhance public protection, to streamline budgeting and administrative tasks, and to continue the use of Integrated Product Teams (IPTs). IPTs are working groups comprised of stakeholders from all levels of government to resolve issues and implement the program.

Within the past year, the program has achieved many notable successes. In a few months we will have made the transition to Acute Exposure Guideline Levels, thus making CSEPP among the first the move to these new and more accurate toxicity values. We have continued the development and installation of automated CSEPP financial management software that automates the application, review, approval, award, reporting, and closeout of CSEPP grants and cooperative agreements. We are developing CSEPP automation tools that seamlessly link to existing commercial emergency management software programs in conformance with common industry standards and protocols. Our emergency exercises are jointly conducted with the full cooperation and involvement of the Army installations and the off-post communities. Army and off-post personnel routinely train together, participate in integrated process teams, and meet to share ideas at meetings and workshops. This relationship has solidified communications between the on-post and off-post communities that is critical in the effective implementation of an emergency plan.

In addition, within the past year we have published a comprehensive Workbook to aid communities in post-emergency recovery planning and are currently developing a computer-based interactive planning and training tool for recovery planning. Our Medical Quality Improvement Teams, relying heavily on state and local expertise, have established Medical Evaluation Guides that serve as a self-evaluation tool for hospitals and Emergency Medical Services organizations to assess their capability to meet their community's needs in the event of any mass casualty situation.

Our Protective Actions IPT is currently working on the development of a Workbook to aid State and local government in the development of protective action decision strategies. We make a concentrated effort to share the results of our research and science with others who can benefit. For example, much that CSEPP has developed can be of value in protection against toxic industrial chemicals, and we have worked closely with the National Institute for Chemical Studies to ensure that CSEPP best practices are exported to other hazardous materials preparedness programs.

At the request of the Alabama community, FEMA has developed an automated protective action decision tool that takes data from the Army's CSEPP hazard analysis model and rapidly develops a course of action to minimize exposure of the population at risk.

CSEPP has continued its history of developing valuable training and public education materials. In early 2003, the CSEPP-developed *Residential Shelter-in-Place* video became a staple for educating the public on protecting themselves from toxic materials, and was made available over the internet and through multiple other sources, in various media. CSEPP is currently developing a companion video on *Shelter-In-Place for Businesses and Industry* for release early in calendar year 2004. Other CSEPP training initiatives during 2003 included development of a video training program for medical professionals entitled *CSEPP Hospital Responses And Protocols*; pilot testing of a new course on a national CSEPP medical curriculum; revision of the popular Joint Information Center Technology course; and an update of the Joint Information Center Advisor on CD-ROM. Since 1999, well over 225,000 CSEPP training courses, job aids, technical reports and public education tools have been downloaded from the CSEPP training web site and are regularly used for training first responders.

Locally, Army and FEMA program personnel worked especially closely with State and county officials in Alabama that resulted in the successful resolution of several emergency preparedness issues. We are in continuing dialogue with our State, County, and Tribal Nation Partners to resolve outstanding issues and to sustain the high level of preparedness that exists within our program.

CONCLUSION

CSEPP has accomplished its mission of protecting the environment, general public, and workers at the eight chemical weapons stockpile sites. Nearly all-major CSEPP systems are in place and operational. Improvements to protections in all communities will continue. Soon all sites will enter the sustainment phase, which will continue until all the chemical weapons stockpiles are destroyed. The sustainment phase does not diminish public protection, but rather ensures that the protections continue.

FEMA is committed to ensuring the successful implementation of CSEPP and protecting the health and safety of our citizens. This has been, and continues to be, a formidable task. As previously stated, "maximum protection" is the most stringent requirement of any emergency preparedness program directive, but we are confident that the Federal, State and local emergency management community is up to the challenge.

There have been significant improvements since 1988 in the capabilities of the CSEPP States and local jurisdictions to respond to a chemical weapons incident. These improvements aid CSEPP communities in protecting their citizens from the full range of natural disasters and technological hazards they face. We are thankful to the CSEPP communities for their commitment and dedication to this important program and pledge to work with them until the risk of a chemical stockpile incident no longer exists. Although FEMA works with State and local government every day to manage risks and mitigate the effects of all kinds of disasters, CSEPP is unique in that the ultimate outcome is the total elimination of the risk to the public from the U.S. Army's chemical stockpile through its destruction. We all look forward to that day when the last chemical weapon and warfare agent is destroyed.

DOCUMENTS SUBMITTED FOR THE RECORD

OCTOBER 30, 2003

September 4, 2003

Acquisition

The Chemical Demilitarization Program: Increased Costs for Stockpile and Non-Stockpile Chemical Materiel Disposal Programs
(D-2003-128)

Department of Defense
Office of the Inspector General

Quality *Integrity* *Accountability*

Additional Copies

To obtain additional copies of this report, visit the Web site of the Inspector General of the Department of Defense at www.dodig.osd.mil/audit/reports or contact the Secondary Reports Distribution Unit of the Audit Followup and Technical Support Directorate at (703) 604-8937 (DSN 664-8937) or fax (703) 604-8932.

Suggestions for Future Audits

To suggest ideas for or to request future audits, contact the Audit Followup and Technical Support Directorate at (703) 604-8940 (DSN 664-8940) or fax (703) 604-8932. Ideas and requests can also be mailed to:

ODIG-AUD (ATTN: AFTS Audit Suggestions)
Inspector General of the Department of Defense
400 Army Navy Drive (Room 801)
Arlington, VA 22202-4704

Defense Hotline

To report fraud, waste, or abuse, contact the Defense Hotline by calling (800) 424-9098; by sending an electronic message to Hotline@dodig.osd.mil; or by writing to the Defense Hotline, The Pentagon, Washington, DC 20301-1900. The identity of each writer and caller is fully protected.

Acronyms

AT&L	Acquisition, Technology, and Logistics
GAO	General Accounting Office

INSPECTOR GENERAL
DEPARTMENT OF DEFENSE
400 ARMY NAVY DRIVE
ARLINGTON, VIRGINIA 22202–4704

September 4, 2003

MEMORANDUM FOR UNDER SECRETARY OF DEFENSE FOR ACQUISITION, TECHNOLOGY, AND LOGISTICS
AUDITOR GENERAL, DEPARTMENT OF THE ARMY

SUBJECT: The Chemical Demilitarization Program: Increased Costs for Stockpile and Non-Stockpile Chemical Materiel Disposal Programs (Report No. D-2003-128)

We are providing this report for your review and comment. This report is the third and last in a series of reports that discuss the management of the Chemical Demilitarization Program. The first report addressed the need to review the acquisition program baseline and to prepare a threat assessment. The second report addressed the improvements that could be made in the oversight, the execution, and the administration of the Chemical Demilitarization Program. This report discusses the factors that continue to affect the cost and schedule of the Chemical Stockpile Disposal Program, and the need to plan for the disposal of buried chemical materiel. We considered management comments on a draft of this report in preparing the final report.

DoD Directive 7650.3 requires that all issues be resolved promptly. In response to the final report, we request that the Under Secretary of Defense for Acquisition, Technology, and Logistics provide comments on Recommendation B.1. by October 6, 2003.

If possible, please provide management comments in electronic format (Adobe Acrobat file only) to audam@dodig.osd.mil. Copies of the management comments must contain the actual signature of the authorizing official. We cannot accept the / Signed / symbol in place of the actual signature. If you arrange to send classified comments electronically, they must be sent over the SECRET Internet Protocol Router Network (SIPRNET).

We appreciate the courtesies extended to the staff. Questions should be directed to Mr. John E. Meling at (703) 604-9091 (DSN 664-9091) or Mr. Rodney D. Britt at (703) 604-9096 (DSN 664-9096). See Appendix E for the report distribution. Audit team members are listed inside the back cover.

Mary L. Ugone
Mary L. Ugone
Deputy Director
Acquisition Management Directorate

Table of Contents

Office of the Inspector General of the Department of Defense

Report No. D-2003-128 **September 4, 2003**
(Project No. D2003AE-0070)

The Chemical Demilitarization Program: Increased Costs for Stockpile and Non-Stockpile Chemical Materiel Disposal Programs

Executive Summary

Who Should Read This Report and Why? Policy makers and milestone decision makers should be interested in this report because it discusses factors that continue to affect the cost and schedule of the Chemical Stockpile Disposal Program and the need to plan for the disposal of non-stockpile chemical materiel.

Background. This report is the third and last in a series of reports that discuss the management of the Chemical Demilitarization Program (the Demilitarization Program). The first report discussed the need for the Army to revise its acquisition program baseline agreement and to obtain a documented threat assessment for the Demilitarization Program. The second report discussed the improvements that could be made in the oversight, the execution, and the administration of the Demilitarization Program. In 1985, the Congress directed DoD to oversee the destruction of the chemical weapons stockpiled munitions and assigned the Army responsibility for the destruction. The Army established the Program Manager for Chemical Demilitarization to manage the day-to-day operations of destroying the chemical weapons. In 1992, the National Defense Authorization Act for FY 1993 (Public Law 102-484) directed the Army to plan for destroying U.S. non-stockpile chemical weapons. In May 2001, the Under Secretary of Defense for Acquisition, Technology, and Logistics (AT&L) designated the Demilitarization Program as a Major Defense Acquisition Program and assigned the Army as the Executive Agent. The Office of the Secretary of Defense approved a life-cycle cost estimate of $24 billion for the Demilitarization Program in September 2001. In February 2003, the Army restructured the program's management by assigning the functions of the Program Manager for Chemical Demilitarization to the Program Manager for the Elimination of Chemical Weapons for plant construction and systemization to the Deputy Director for Plant Operations for operations and closure. The restructuring also assigned the Director, Chemical Materials Agency to manage the overall Demilitarization Program. Through May 2003, the Army awarded contracts totaling $5.7 billion for the construction, systemization, operations, and closure of seven chemical agent disposal facilities and planned two additional disposal facilities.

Results. The Director, Chemical Materials Agency had made substantial progress in managing the cost growth for the Demilitarization Program; however, several issues could affect the future program cost and schedule of the Chemical Stockpile Disposal Program and the disposal of the non-stockpile chemical materiel. Specifically:

- The Director, Chemical Materials Agency's ability to effectively control the cost estimate of the Chemical Stockpile Disposal Program continues to be affected by delays in obtaining State permit modifications needed for beginning disposal operations, monetary effects of decisions on the type of technology to be employed at two Assembled Chemical Weapons Assessment facilities, the escalation in costs and safety incidents at operational chemical disposal facilities, and rising cost estimates for closure of disposal facilities (finding A).

- The Product Manager for Non-Stockpile Chemical Materiel did not have information needed to prepare a reliable estimate of the cost and schedule to dispose of buried chemical warfare materiel. Direction to the environmental offices of the DoD Components will cause DoD Components to identify, schedule, and fund the disposal of buried chemical warfare materiel from existing and former DoD installations. Implementation of the direction will also result in a reliable and defendable estimate of the cost to dispose of the buried chemical warfare materiel for the contingent liability in Note 16 of the DoD financial statements (finding B).

For details of the audit results, see the Findings section of the report.

Management Comments. The Deputy Assistant to the Secretary of Defense (Chemical Demilitarization and Threat Reduction), Office of the Assistant to the Secretary of Defense (Nuclear and Chemical and Biological Defense Programs), responding for the Under Secretary of Defense for Acquisition, Technology, and Logistics, suggested some editorial changes to the report that we considered and made where deemed appropriate. However, the Deputy Assistant did not respond to the recommendation to issue direction to the environmental offices of the DoD Components to identify, schedule, and fund the disposal of buried chemical warfare materiel. The Director, U.S. Army Chemical Materials Agency concurred with the recommendation for the Product Manager for Non-Stockpile Chemical Materiel to prepare updated cost estimates for all burial sites. In response to the final report, we request that the Under Secretary of Defense for Acquisition, Technology, and Logistics comment on the need to address planning for the disposal of buried chemical warfare materiel by October 6, 2003.

Background

This report is the third and last in a series of reports that address the management of the Chemical Demilitarization Program (the Demilitarization Program). The first report discussed the need for the Army to revise its acquisition program baseline agreement and to obtain a documented threat assessment for the Demilitarization Program. The second report discussed improvements that could be made in the oversight, the execution, and the administration of the Demilitarization Program. This report discusses the key factors that continue to affect the cost and schedule of the Chemical Stockpile Disposal Program and the need to plan for the disposal of buried chemical warfare materiel. The mission of the Demilitarization Program is to develop and operate facilities that destroy chemical munitions through incineration and other approved alternative destruction technologies.

Program History. In 1985, Congress compelled the DoD to establish the Demilitarization Program. Specifically, due to congressional concerns for the stockpile's deterioration, section 1521, title 50, United States Code, "Destruction of Existing Stockpile of Lethal Chemical Agents and Munitions," (Public Law 99-145), directed DoD to oversee the destruction of the stockpile of lethal chemical agents and munitions. The Congress, as part of the same legislation, designated the Army as the Military Department responsible for the destruction of the stockpile. Later in 1985, the Army assigned the Program Manager for Chemical Demilitarization to manage the day-to-day operations of destroying the chemical munitions. In 1992, the National Defense Authorization Act for FY 1993 (Public Law 102-484) directed the Army to plan for destroying U.S. non-stockpile chemical weapons. In May 2001, the Under Secretary of Defense (AT&L) designated the Demilitarization Program as a Major Defense Acquisition Program (Acquisition Category ID), with the Army as the Executive Agent. In May 2002, the Under Secretary of Defense (AT&L) certified to Congress, pursuant to section 2433, title 10, United States Code, "Unit Cost Reports," (Public Law 99-500), that the Demilitarization Program was essential to national security, that no alternatives existed to the program, that new cost estimates were reasonable, and that management was adequate to control program costs.

Program Management and Status. The Assistant to the Secretary of Defense (Nuclear and Chemical and Biological Defense Programs) oversees the efforts of the Demilitarization Program for the Under Secretary of Defense (AT&L). The Assistant Secretary of the Army (Acquisition, Logistics, and Technology) oversees planning, programming, and budgeting for the Demilitarization Program. The Commanding General, Army Materiel Command oversees the chemical disposal operations and emergency preparedness. The Demilitarization Program consists of five individual programs: Chemical Stockpile Disposal, Non-Stockpile Chemical Materiel, Alternative Technologies and Approaches, Chemical Stockpile Emergency Preparedness, and Assembled Chemical Weapons Assessment.

In February 2003, the Army restructured the program's management by establishing the Director, Chemical Materials Agency to manage the overall Demilitarization Program and assigning the functions that the Program Manager for Chemical Demilitarization performed on plant construction and systemization to the Program Manager for the Elimination of Chemical Weapons and operations and closure to the Deputy Director for Plant Operations. The Project Managers for the Chemical Stockpile Disposal, the Alternative Technologies and Approaches, and the Non-Stockpile Chemical Materiel report to the Program Manager for the Elimination of Chemical Weapons. The Project Manager for Chemical Stockpile Emergency Preparedness reports to the Deputy for Plant Operations, Office of the Director, Chemical Materials Agency, and the Program Manager for Assembled Chemical Weapons Assessment reports to the Under Secretary of Defense (AT&L). The mission of the Chemical Stockpile Disposal Project is to destroy the U.S. stockpile of unitary chemical agents and munitions. The missions of the Alternative Technologies and Approaches Project and the Chemical Stockpile Emergency Preparedness Program support the Chemical Stockpile Disposal Project and the Non-Stockpile Chemical Materiel Product through examining and demonstrating alternative destruction technologies and enhancing protection of the public, the workers, and the environment, respectively. The Assembled Chemical Weapons Assessment Program has the mission to validate, demonstrate, and implement alternative destruction technologies for assembled chemical weapons at Pueblo, Colorado, and Blue Grass, Kentucky. The mission of the Non-Stockpile Chemical Materiel Product is to destroy all binary chemical munitions, former chemical weapon production facilities, recovered chemical warfare materiel, and miscellaneous warfare materiel. Appendix C describes the existing management roles and the revised management roles within the Demilitarization Program in more detail.

Program Cost Estimates and Contract Awards. In September 2001, the Under Secretary of Defense (AT&L) approved a cost estimate prepared by the Cost Analysis Improvement Group, Office of the Secretary of Defense, which increased the total estimated program cost to $24 billion. Through May 2003, the Army awarded contracts totaling $5.7 billion for seven chemical agent disposal facilities. Specifically, the U. S. Army Corps of Engineers awarded contracts for the construction, systemization, operations, and closure of disposal facilities at Johnston Island, Hawaii, and Tooele, Utah, the two facilities that had reached or passed the operational phase. The U.S. Joint Munitions Command had also awarded contracts for five other disposal facilities. The U.S. Joint Munitions Command also awarded contracts for the other demilitarization programs. DoD funds the Demilitarization Program through the Chemical Agents and Munitions Destruction Account, which includes Military Construction.

Objectives

The overall objective was to evaluate the overall management of the Demilitarization Program and associated management controls. Specifically, we evaluated the Director's efforts to contain cost growth within the Chemical Stockpile Disposal Program and DoD plans to destroy U.S. non-stockpile chemical weapons. Appendix A discusses the results of the review of management controls and the scope and methodology of the review. Appendix B identifies prior audit coverage of the Demilitarization Program.

A. Key Factors Continue to Affect the Cost Estimate for the Chemical Stockpile Disposal Program

Many key factors continue to affect the Director, Chemical Materials Agency's (the Director) ability to effectively control costs for the Chemical Stockpile Disposal Program. Specifically, the Director has been affected by costly delays in reaching public consensus when obtaining State permit modifications needed to begin disposal operations, monetary effects of decisions on the type of technology to be employed at two Assembled Chemical Weapons Assessment facilities, the cost escalation and safety incidents at operational chemical disposal facilities, and rising cost estimates for closure of disposal facilities. These conditions exist because the Chemical Demilitarization Program (the Demilitarization Program) is a very large and complex program influenced by several offices within and outside of the Department of Defense. As a result, the Army will continue to experience cost growth in funds needed to complete Demilitarization Program requirements. This program cost growth may also lead to additional program baseline cost breaches that will require the Under Secretary of Defense (AT&L) to again certify the program's cost and schedule to the Congress.

Earlier Audit Reports on Factors Affecting Program Costs

The General Accounting Office (GAO) has previously reported on issues affecting the Demilitarization Program's cost and schedule. In Report No. NSIAD 97-18, "Chemical Weapons and Materiel: Key Factors Affecting Disposal Costs and Schedule," February 10, 1997, GAO reported that key factors affecting the program's cost and schedule included public concerns over the safety of incineration, legislative requirements, the introduction of alternative disposal technologies, and compliance with environmental laws and regulations.

Safety of Incineration Facilities. In 1988, the public voiced concern over the Army's decision to use on-site incineration to dispose of the chemical warfare materiel. The Congress responded to those concerns by directing the Army, through legislative requirements, to assess and report on potential alternative technologies to incineration

Legislative Requirements. Since 1985, when the Congress directed the Army to destroy the U.S. stockpile of chemical materiel, other legislative requirements that limited disposal options have affected the Army's ability to control the program's cost and schedule. The program limitations made by the Congress included "The National Defense Authorization Act for Fiscal Year 1991" (Public Law 101-510), which restricted DoD from using funds to transport chemical weapons to Johnston Atoll except for U.S. munitions discovered in the Pacific, and also restricted DoD from studying the movement of chemical munitions. Additionally, "The National Defense Authorization Act for Fiscal Year 1993" (Public Law 102-484) directed the Army to study alternatives to incineration.

Alternative Disposal Technologies. In November 1991, because of public concern and congressional direction, the Army requested the National Research Council to evaluate potential technological alternatives to the baseline incineration process. In Public

Law 102-484, the Congress directed the Army to consider using the potential technological alternatives to incineration that were identified in the National Research Council's report. As a result, in 1994, the Army initiated the Alternative Technologies and Approaches Project at the two bulk-only chemical stockpile sites, Aberdeen Proving Ground, Maryland, and Newport Chemical Activity, Indiana, to investigate, develop, and support testing of two technologies based on neutralization of chemical agents. In "The National Defense Authorization Act for Fiscal Year 1997" (Public Law 104-201), the Congress directed DoD to conduct an assessment of alternative technologies for the disposal of assembled chemical munitions. Additionally, in the 1997 Appropriations Act, the Congress prohibited the Army from obligating funds for constructing disposal facilities at Blue Grass, Kentucky, and Pueblo, Colorado, until 180 days after the Secretary of Defense reported on the alternative technologies.

Compliance with Environmental Laws. Before constructing or operating a chemical disposal facility, the Army must obtain permits to comply with Federal, State, and local environmental laws and regulations. The Resource Conservation and Recovery Act regulates the storage, treatment, and disposal of most chemical materiel. The Act controls hazardous waste through a permit process that requires Government approval for the generation, transportation, storage, and disposal of hazardous waste. Additionally, the Act allows the Environmental Protection Agency to authorize individual States to administer and enforce hazardous waste programs. Under the Act, the States can establish programs that are more stringent than the Federal program. The GAO believed that the permit process would take more time than the Army allowed in its schedule.

Complex Management Structure. In Report No. NSIAD 00-80, "Chemical Weapons Disposal: Improvements Needed in Program Accountability and Financial Management," May 8, 2000, GAO stated that effective management of the Demilitarization Program was hindered by its complex management structure and ineffective coordination among program offices and State and local officials. Specifically, the GAO reported that several changes in the organization and structure of the program from 1997 through 1999, including some changes to implement legislative requirements, divided the management roles, responsibilities, and accountability among several different management levels within the DoD and the Army. As the program expanded beyond its original single purpose of destroying the stockpile to encompass a broader range of missions, to include compliance with the Chemical Weapons Convention, the organization and structure of the program became increasingly complex.

GAO further reported that, at times, the several different levels within DoD and the Army shared oversight responsibilities, resulting in fragmented responsibilities for management decisions. The fragmented management affected the ability of DoD and the Army to present a coordinated message to State and local officials for the Blue Grass, Kentucky, and Pueblo, Colorado, stockpile sites. The GAO reported that the confusion at these two sites led to the public's perception that the program lacked a single vision for destroying the chemical stockpile in a judicious manner.

Increase in the Demilitarization Program's Cost Estimate

In September 2001, the Under Secretary of Defense (AT&L) approved a revised cost estimate for the Demilitarization Program. The revised cost estimate of $24 billion was substantially more than the cost estimate of $15.3 billion approved in 1998. The Office of the Cost Analysis Improvement Group advised that the increase was necessary to reflect more realistic destruction rates than had been previously estimated. Table 1 shows the DoD-approved program cost estimate as of September 2001.

Table 1. Approved Cost Estimate for the Demilitarization Program - September 2001

Program/Facility	Estimated Cost ($ in millions)
Chemical Stockpile Disposal Project	$ 3,086
Johnston Atoll	1,830
Tooele	2,364
Anniston	2,298
Pine Bluff	1,737
Umatilla	2,462
Subtotal	13,777
Alternative Technologies and Approaches Project	177
Aberdeen	1,019
Newport	1,459
Subtotal	2,655
Assembled Chemical Weapons Assessment Program[1]	0
Pueblo [2]	1,784
Blue Grass [3]	2,072
Subtotal	3,856
Non-Stockpile Chemical Materiel Product	1,632
Chemical Stockpile Emergency Preparedness Program	2,139
Total	**$ 24,059**

[1] Program costs for Assembled Chemical Weapons Assessment were zero because Pueblo and Blue Grass were part of the Chemical Stockpile Disposal Project during the September 2001 Defense Acquisition Board review.
[2] The Pueblo cost estimate reflects modified baseline incineration costs and was included in costs for the Chemical Stockpile Disposal Project.
[3] The Blue Grass cost estimate reflects baseline incineration costs and was included in costs for the Chemical Stockpile Disposal Project.

In February 2003, the Director presented a revised program office estimate to the Army Cost and Economic Analysis Center and the Army Cost Review Board. The Director prepared the revised cost estimate to document the changes in the program since the September 2001 cost estimate and in support of the program objective memorandum for

FY 2005. Table 2 shows the Office of the Chemical Materials Agency's revised program cost estimate that was presented to the Army Cost and Economic Analysis Center and the Army Cost Review Board in February 2003.

Table 2. Recommended Cost Estimate for the Demilitarization Program - February 2003

Program/Facility	Estimated Cost ($ in millions)
Chemical Stockpile Disposal Project	$ 2,907
Johnston Atoll	1,761
Tooele	2,395
Anniston	2,426
Pine Bluff	1,974
Umatilla	2,757
Subtotal	14,220
Alternative Technologies and Approaches Project	147
Aberdeen	850
Newport	1,220
Subtotal	2,217
Assembled Chemical Weapons Assessment Program[1]	354
Pueblo	1,537
Blue Grass	2,396
Subtotal	4,287
Non-Stockpile Chemical Materiel Product	1,586
Chemical Stockpile Emergency Preparedness Program	2,809
Total	**$ 25,119**

[1] The Assembled Chemical Weapons Assessment cost estimate was not presented to the Army Cost and Economic Analysis Center or the Cost Review Board for review.

Factors Affecting Program Costs Continue

Despite Army efforts to contain the cost growth of the Demilitarization Program, factors similar to those previously reported by the GAO continue to affect the ability of the Director, Chemical Materials Agency to effectively control program costs. Specifically, the Director has been affected by costly delays in reaching public consensus with obtaining State permit modifications needed for beginning disposal operations, the decisions on the type of technology to be employed at two Assembled Chemical Weapons Assessment facilities, the escalation of costs and safety incidents at operational chemical disposal facilities, and rising cost estimates for closure of disposal facilities.

Environmental Permits. The Director, Chemical Materials Agency needs to receive public agreement from the involved States when an environmental permit modification is required before the Army will approve the start of disposal operations. The process of receiving public agreement with the environmental permit process continues to be a major roadblock in containing costs. To illustrate, the chemical disposal facility in Anniston, Alabama, completed surrogate (agent trial) testing in January 2003 and was approved, subject to obtaining the State environmental permit modification, for an agent trial burn plan to begin disposal operations. Because State and local officials disagreed with DoD and Army officials on the level of preparedness needed by residents surrounding the facility, the State of Alabama refused to approve the environmental permit modification, which delayed the start of full disposal operations. According to the Office of the Chemical Stockpile Emergency Preparedness Program, local officials from Calhoun County, Alabama, remain dissatisfied with the level of emergency preparedness provided to residents in the area immediately surrounding the disposal facility. To overcome this dissatisfaction, the State of Alabama requested that the Army provide an additional $26.9 million in FY 2003 to over-pressurize county schools and fully implement a plan to provide shelters in-place for the local residents identified with having special needs. In FY 2002, the Under Secretary of Defense (AT&L) approved and the Army provided to the Federal Emergency Management Agency funding to satisfy a State of Alabama request to provide Calhoun County with $40.5 million in FY 2002 to provide the local residents with escape hoods and updated emergency radios. In August 2003, the Army started limited operations during night and weekend hours to destroy M-55 rockets filled with nerve agent. However, until the schools can be over-pressurized and the special needs for some local residents can be met with the additional $26.9 million in funding, the State of Alabama's environmental office will not agree with the Army's plan to begin full facility disposal operations. In the meantime, the Anniston disposal facility is fully staffed and ready for full operation. The Director, Chemical Materials Agency, in preparing the program cost estimate, estimates that operation and disposal costs are approximately $287,000[1] a day, regardless of whether or not the facility is operational. Delays in receiving timely public agreement to obtaining State environmental permit modifications for other chemical agent disposal facilities could further affect estimated program costs.

Recent Technology Decisions for Assembled Chemical Weapons Assessment Programs. The "National Defense Appropriations Act for Fiscal Year 1997," September 10, 1996 (Public Law 104-208), established the Assembled Chemical Weapons Assessment Program as a separate and distinct program under the Demilitarization Program to research alternative chemical munitions destruction technology for the planned chemical disposal programs at Pueblo, Colorado, and Blue Grass, Kentucky. The Under Secretary of Defense (AT&L) recently made decisions on the type of technology that will be used at the two disposal sites. In July 2002, the Under Secretary of Defense (AT&L) approved neutralization followed by biological treatment[2] as the technology to dispose of chemical weapons at Pueblo, Colorado. In February 2003, the Under Secretary of Defense (AT&L) approved neutralization followed by supercritical water oxidation as the technology to pilot test for the disposal of chemical weapons at Blue Grass, Kentucky.

[1] The daily cost of $287,000 is a rough order magnitude amount that the Program Manager estimated by dividing the total annual chemical disposal facility contract for Anniston by 365 days.

[2] The process of mixing hot water with the chemical agent to a point where it is broken down into other chemical components that can be further treated with bacteria.

At the time that the September 2001 program cost estimate was prepared, the Director, Chemical Materials Agency based the cost estimate for the Pueblo and Blue Grass disposal facilities on the Army's Chemical Materials Agency employing the incineration technology. In June 2002 and November 2002, respectively, the Assembled Chemical Weapons Assessment Program Manager submitted the cost estimate for the Pueblo and Blue Grass disposal facilities based on preliminary disposal facility designs to the Cost Analysis Improvement Group for review and approval. Consequently, the Office of the Cost Analysis Improvement Group assessed the program estimate and provided a revision to the cost estimates for the two disposal facilities. However, the full cost may escalate once the contractors complete the final disposal facility designs.

Contractor Costs. The Director, Chemical Materials Agency has contracts with contractors at eight of the nine chemical disposal facilities to design and construct the facilities, as well as operate the facilities once they are operational. Contractors involved in the design, construction, and operation of disposal facilities have also experienced increased program costs.

Facility Development. Costs to design and construct disposal facilities have also deviated substantially in the past. To illustrate, in October 2001, after the September 2001 program cost estimate was approved, the Director of Contracts issued a modification to the contract for the design, construction, and operation of the Newport Chemical Agent Disposal Facility that increased the contract value from $296 million to $748 million. The Director of Contracts awarded the modification because of the contractor's inability to meet cost and schedule goals. Since the modification's award, the Under Secretary of Defense (AT&L) approved a plan to accelerate the facility's disposal schedule that decreased the contract scope from $748 million to $584 million.

Operation. Chemical agent exposure incidents at the chemical disposal facilities during disposal operations can also significantly affect future program costs. On July 15, 2002, a chemical agent exposure incident occurred at the Tooele Chemical Agent Disposal Facility in Utah. As a result, the Army postponed disposal operations until the contractor completed corrective actions identified by an Army investigation team and the State of Utah agreed that it is safe to restart disposal operations. The operations and maintenance costs for the Tooele, Utah, disposal facility is estimated to cost $336,000[3] a day, regardless of whether or not the facility is operating. On March 28, 2003, the Tooele, Utah disposal facility restarted disposal operations. As of March 27, 2003, disposal operations at the Tooele facility had been idle for 256 days, at a program cost of approximately $86 million.

Disposal Facility Closure Cost Estimates. Cost estimates for closure of disposal facilities at the completion of facility operations had not been fully established. Facility closure costs include the management, expertise, and labor to decommission and close the facilities in accordance with applicable laws and regulations. Closure of the disposal facilities commences at the completion of chemical disposal operations and ends when the contractor fulfills State requirements. The Program Manager for Chemical Demilitarization, in the 1998 cost estimate, planned for 12 months to close each disposal facility. In the September 2001 cost estimate, the Director considered additional costs to close each planned facility but the closure costs still had not been fully defined. Closure

[3] The daily cost of $336,000 is a rough order magnitude amount that the Program Manager estimated by dividing the total annual chemical disposal facility contract at the Tooele facility by 365 days.

costs at the Johnston Atoll Chemical Agent Disposal System, for example, were initially estimated to be $158 million over a period of 12 months. In the September 2001 cost estimate the Cost Analysis Improvement Group revised the estimated cost and schedule to $411 million over 33 months. Through April 2003, the contractor incurred costs of $262 million over 25 months but, because of funding reductions, will not meet the estimated 33-month schedule to complete closure of the facility.

The Director and the Assembled Chemical Weapons Assessments Program Manager indicated that they would negotiate closure requirements for each disposal facility with State environmental offices based on lessons learned from closure activities at the Johnston Atoll Chemical Agent Disposal System. Accordingly, the reasonableness of estimated facility closure costs included in the September 2001 program cost estimate is subject to revision, based on the lessons learned from the Johnston Atoll facility and negotiations with each State environmental office on facility closure requirements.

Influences on the Chemical Demilitarization Program

The Demilitarization Program is a very large and complex program that has been influenced by several offices within and outside the Department of Defense. Many of the issues affecting the program cost and schedule have resulted from the complex program structure and coordination requirements affecting the program management. The primary offices that affect the program's cost and schedule are State and local Governments, special interest groups that challenge the Army's technology decisions, and the complex management structure within DoD that makes key program decisions.

State and Local Governments. State and local Governments play a key role in public officials' opinions and can affect the timeliness of permit decisions. Before granting required major modifications to existing permits, the State environmental regulatory agency typically seeks the general public's input. Public comments can include such issues as the off site emergency preparedness for the communities. In the case of the Anniston, Alabama disposal facility, the surrounding local officials have asserted that the off-site emergency preparedness was not adequate to support the start of operations. Because of State and local concerns and start up requirements articulated by Senator Richard Shelby concerning off-site emergency preparedness addressed in the Chemical Stockpile Emergency Preparedness Project, the Alabama Department of Environmental Management is still evaluating the public input associated with these plans. The representative from the Chemical Stockpile Emergency Preparedness Program stated that coordination with other State and local government officials could potentially result in similar operational delays for other chemical agent disposal facilities.

Special Interest Groups. Special interest groups also influence program functions and could eventually affect the program's cost and schedule. One lawsuit filed in Umatilla, Oregon, by a special interest coalition alleges that the issuance of the Umatilla Chemical Agent Disposal Facility permit on the basis of the findings and conclusions of the Oregon Department of Environmental Quality were not supported by substantial evidence or the State environmental office failed to comply with State and Federal requirements. The special interest coalition includes the Group Against Social Predation, a Hermiston-based opposition group affiliated with the Chemical Weapons Working Group, the Sierra Club, and 22 individual petitioners. In December 2002, the Director reported to the Under Secretary of Defense (AT&L) that the court trial could potentially delay the start of disposal operations at the Umatilla facility by as much as 3 years if the judge ruled in favor of the group. The Chemical Weapons Working Group and the Sierra Club were also responsible for making similar allegations at other disposal facilities.

Management Structure. The complex management structure that oversees the Demilitarization Program could also affect the program cost and schedule. As noted in the GAO reports, the complex program structure had hindered program management. The issue noted by GAO still exists as management and oversight of the Demilitarization Program is still evolving. Specifically, in February 2003, after the chemical exposure incident at the Tooele, Utah, disposal facility, the Army restructured the management of the Demilitarization Program to align acquisition oversight, previously under the Assistant Secretary of the Army (Installations and Environment), under the Assistant Secretary of the Army (Acquisition, Logistics, and Technology). At the same time, the Army placed the management of the operation of the chemical disposal facilities, previously under the Program Manager for Chemical Demilitarization, under the Commanding General, Army Materiel Command. The Army believes that the realignment of functions and responsibilities will provide better oversight of the program acquisition process and of the contractors' chemical surety management[4] at the Government-owned disposal facilities.

Conclusion

Future cost growth of the Demilitarization Program seems likely. As a result, the Army will continue to experience cost growth in funds needed to complete Demilitarization Program requirements. This program cost growth may also lead to additional program baseline cost breaches that will require the Under Secretary of Defense (AT&L) to recertify the program's cost and schedule to the Congress.

Management Comments on the Finding and Audit Response

The detailed responses to the comments from the Acting Deputy Assistant Secretary of the Army for the Elimination of Chemical Weapons, Office of the Assistant Secretary of the Army (Acquisition, Logistics, and Technology) follow. The Acting Deputy Assistant Secretary commented on the draft report's discussion on estimating the cost and schedule for disposal facility closure and additional funding for emergency preparedness in the State of Alabama. The complete text of those comments is in the Management Comments section of this report.

Disposal Facility Closure Cost and Schedule Estimates. The Acting Deputy Assistant Secretary recommended that we reexamine statements made about disposal facility closure costs. The Acting Deputy stated that cost estimates for the closure of incineration-based disposal facilities are not expected to negatively affect future program cost estimates because the 33-month duration, $411 million closure cost, estimated in September 2001 for the Johnston Island, Hawaii, disposal facility, will be accomplished in 31 months and cost $365 million.

Audit Response. The Director, U.S. Army Chemical Materials Agency did successfully close the Johnston Island, Hawaii, disposal facility earlier and at a lower cost than had

[4] Oversight of the chemical disposal facility operations includes testing, maintenance, and safety.

been planned in September 2001. Closure costs, however, will continue to be a key factor that affects the program life cycle costs until each State environmental office with a disposal facility site identifies its specific facility closure requirements. Until the State environmental offices identify their closure requirements, the Program Manager cannot make definitive estimates of the closure costs and the schedule for each of the disposal sites.

State of Alabama Emergency Preparedness Funding. The Acting Deputy Assistant Secretary stated that the Army provided the additional FY 2003 funding for collective protection and special population requirements identified by the State of Alabama.

Audit Response. We revised the report to acknowledge that the Army had provided the additional funds to satisfy the State of Alabama's request.

B. Disposal of Buried Chemical Warfare Materiel

The Product Manager for Non-Stockpile Chemical Materiel did not have information needed to prepare a reliable estimate of the cost and schedule to dispose of buried chemical warfare materiel. This condition occurred because the Under Secretary of Defense (AT&L) had not directed the DoD Components to identify, schedule, and fund the disposal of buried chemical warfare materiel from existing and former DoD installations. As a result, the Product Manager for Non-Stockpile Chemical Materiel was unable to fully satisfy the congressional direction to provide an actionable plan for disposal of all non-stockpile chemical warfare materiel. Also, without an actionable plan, the Under Secretary of Defense (AT&L) cannot inform the Congress and the public of the realistic costs and the planned schedule to dispose of buried chemical warfare materiel. Furthermore, the Product Manager cannot replace the $8.9 billion contingent liability, which was prepared as a rough order magnitude estimate in Note 16 of the DoD financial statements, with a reliable and defendable estimate of the cost to dispose of the buried chemical warfare materiel.

Policy on Non-Stockpile Chemical Materiel Disposal

Public Law 102-484. In the National Defense Authorization Act for Fiscal Year 1993, section 176, "Report on Destruction of Non-Stockpile Chemical Materiel," (Public Law 102-484), Congress directed the Secretary of the Army to submit a report not later than February 1, 1993, that provided a plan for the remediation[5] of all chemical warfare material of the United States not covered by section 1521, title 50, United States Code, "Destruction of Existing Stockpile of Lethal Chemical Agents and Munitions," (Public Law 99-145). The Congress further directed that the report identify the locations, types, and quantities of non-stockpile chemical materiel, explain the methods to be used for their disposal, provide the estimated cost and schedule for their disposal, and discuss transportation alternatives.

Defense Environmental Restoration Program. The "Management Guidance for the Defense Environmental Restoration Program," September 2001, states that goals of the Defense Environmental Restoration Program include the identification, investigation, research and development, and cleanup of contamination from hazardous substances, pollutants, and contaminants, and the correction of other environmental damage (such as detection and disposal of unexploded ordnance) that creates an imminent and substantial endangerment to the public health or welfare or to the environment.

Army Policy. Army Regulation 200-1, "Environmental Protection and Enhancement," February 21, 1997, implements the Army strategy to restore previously contaminated sites that pose a threat to human health and the environment. Through the Regulation, the Army provides policy for the Army's Environmental Restoration Programs, including the Installation Restoration Program for real property that is controlled by the active

[5] According to the Environmental Planning and Assessment Act of 1979, remediation is defined as removing, dispersing, destroying, reducing, mitigating or containing the contamination of any land.

Army installations; formerly used defense sites including real property that was owned by, leased to, possessed by, or otherwise under the operational control of the Secretary of Defense or other Military Components that predated the Department of Defense; and the Base Realignment and Closure Program.

Financial Management Regulation. DoD Regulation 7000.14-R, "DoD Financial Management Regulation," Volume 4, Chapter 12, November 1999, states that contingencies are existing conditions, situations, or circumstances involving uncertainty as to possible gain or loss to an entity. A loss contingency exists when the likelihood that the future event or events will confirm the loss or impairment of an asset or the incurrence of a liability can be classified as probable, reasonably possible, or remote. Contingent liabilities should be recorded in DoD financial systems and reported in financial statements when a future outflow or other sacrifice of resources is probable or measurable, and disclosure is necessary if the financial statements would otherwise be misleading.

Estimated Cost to Dispose of Buried Chemical Warfare Materiel

GAO Report No. NSIAD 97-18, "Chemical Weapons and Materiel: Key Factors Affecting Disposal Costs and Schedule," February 10, 1997, reported that the Non-Stockpile Chemical Materiel Product would need $14.5 billion to dispose of all buried chemical warfare materiel, which was 95 percent of the total Non-Stockpile Program cost estimate of $15.2 billion. According to a representative from the Non-Stockpile Product Office, discussions above the Department of the Army level resulted in excluding the disposal of buried chemical warfare materiel from the Demilitarization Program because costs were high. Accordingly, the February 2003, cost estimate for the Non-Stockpile Chemical Materiel Product included only $1.586 billion to dispose of non-stockpile chemical warfare materiel declared under the Chemical Weapons Convention and to continue research, development, and testing of non-stockpile chemical warfare disposal technologies. To meet the congressional requirement to plan for the disposal of the buried munitions, the Non-Stockpile Chemical Materiel Product Manager estimated that the mission would cost an additional $11.7 billion.[6]

Implementation of Actions to Dispose of Buried Chemical Warfare Materiel

The Product Manager for Non-Stockpile Chemical Materiel did not have information needed to prepare a reliable estimate of the costs and schedule to dispose of buried chemical warfare materiel with DoD activities. A key piece of information necessary is a preliminary assessment of the intended use of the installations to determine the extent that the DoD will go through remediation. As discussed below, environmental restoration programs have been established, remediation efforts have begun, yet

[6] The $11.7 billion estimate was based on the rough order magnitude estimate developed in 1993; refined in 1996 using the most up-to-date information on burial site characterization, destruction technology decisions, and remediation and treatment procedures; and adjusted in 2002 using inflation indexes. The $11.7 billion estimate is in current year dollars and is equivalent to the $8.9 billion estimate reported in Note 16 of the financial statements, which is in base year dollars.

additional action needs to be taken to identify, schedule, and fund the disposal of recovered buried chemical warfare materiel from existing and former DoD installations.

Environmental Restoration Programs. The "Survey and Analysis Report, Second Edition," December 1996, states that the continental United States, the U.S. Virgin Islands, and the District of Columbia contain 224 installations where chemical warfare materiel may be buried. Preliminary analysis in the report stated that 56 of the 224 installations (25 percent) may require no further action and 168 of the 224 installations (75 percent) may require remediation. The Army established environmental restoration programs to evaluate the need and execute the removal of the chemical warfare materiel from those burial sites. The 224 installations include active, formerly used defense sites, and base realignment and closure sites. See Appendix D for more information regarding the environmental restoration programs and Army roles and responsibilities within each program.

Implementation of Remediation Efforts. The Under Secretary of Defense (AT&L) had not issued direction requiring the DoD Components to develop and prioritize a destruction schedule for remediation at all chemical warfare material burial sites. As Executive Agent, the Secretary of the Army did assign responsibility for the cleanup of formerly used defense sites under the Defense Environmental Restoration Program to the U.S. Army Corps of Engineers (Army Corps). The Army Corps stated that the process for cleaning up the formerly used sites where chemical warfare materiel may be located was in its initial phase. Specifically, the Army Corps was performing site surveys to determine the scope and magnitude of seven burial sites. The Army Corps plans to report the results of their survey to the Assistant Secretary of the Army (Environment Safety and Occupational Health) by September 2004. Although the Army Corps will address the formerly used defense sites, the Under Secretary of Defense (AT&L) had not tasked the other DoD Components to perform a similar assessment for sites located on active installations and base realignment and closure installations previously identified.

Actions Needed. During the review, a representative from the Office of the Deputy Under Secretary of Defense (Installations and Environment) agreed that the DoD needed to establish top-level direction to:

- schedule the remediation of all potential chemical warfare materiel burial sites; and,

- assign DoD Components with responsibility for prioritizing a schedule for surveying and excavating the sites that potentially have buried chemical warfare materiel.

A destruction schedule that prioritizes the remediation of chemical warfare material burial sites needs to be established so that the DoD Components will plan and estimate costs for excavation, removal, destruction, and treatment procedures for each burial site. Because the Under Secretary of Defense (AT&L) had not issued direction to the DoD Components, the Product Manager for Non-Stockpile Chemical Materiel has not been able to fully identify the funding requirements needed to satisfy the Congressional direction to plan for the disposal of all chemical warfare materiel. Additionally, the Under Secretary of Defense (AT&L) cannot make the Congress and the public timely aware of the costs and schedules to dispose of buried chemical warfare materiel. Furthermore, the Product Manager cannot replace the $8.9 billion contingent liability, which was prepared as a rough order magnitude estimate in Note 16 of the DoD financial statements, with a reliable and defendable estimate of the cost to dispose of the buried chemical warfare materiel.

Management Comments on the Finding and Audit Response

The detailed responses on the comments from the Acting Deputy Assistant Secretary of the Army for the Elimination of Chemical Weapons, Office of the Assistant Secretary of the Army (Acquisition, Logistics, and Technology) responding for the Director, U.S. Army Chemical Materials Agency follow. The Acting Deputy Assistant Secretary commented on the draft report's statement on the cost estimate for the disposal of buried munitions. The complete text of those comments is in the Management Comments section of this report.

The Cost Estimate for Buried Chemical Warfare Materiel. The Acting Deputy Assistant Secretary commented on the $11.7 billion cost estimate for disposal of buried chemical warfare materiel. The Acting Deputy stated that the estimate had not been updated since 1996 except for an adjustment in the inflation indices. The Acting Deputy stated a new cost estimate using current remediation, treatment procedures, technology, site information, and environmental standards will significantly increase the cost estimate.

Audit Response. We agree with the Acting Deputy Assistant Secretary's comments.

Recommendations, Management Comments, and Audit Response

B.1. We recommend that the Under Secretary of Defense for Acquisition, Technology, and Logistics issue direction to the environmental offices of the DoD Components to identify, schedule, and fund the disposal of buried chemical warfare materiel from active installations and base realignment and closure installations previously identified.

Management Comments. The Under Secretary of Defense for Acquisition, Technology, and Logistics did not respond to the recommendation. In response to the final report, we request that the Under Secretary comment on the need to issue direction to identify, schedule, and fund the disposal of buried chemical warfare materiel from active installations and base and realignment and closure installations previously identified.

B.2. We recommend that the Product Manager for Non-Stockpile Chemical Materiel update the plan and cost estimate for disposal of buried munitions after the environmental offices of the DoD Components implement Recommendation B.1.

Acting Deputy Assistant Secretary of the Army for the Elimination of Chemical Weapons and Director, U.S. Army Chemical Materials Agency Comments. The U.S. Army Chemical Materials Agency concurred, stating that the Product Manager for Non-Stockpile Chemical Materiel is preparing cost estimates for four potential base and realignment and closure installations and is prepared to update the estimates for all burial sites. Further, during the fourth quarter of FY 2003, the Product Manager will meet with representatives from the Office of the Under Secretary of Defense for Acquisition, Technology, and Logistics to discuss burials and the path forward.

The Acting Deputy nonconcurred with Recommendation B.2. as written, stating that the Product Manager should participate as a subject-matter expert in updating the cost estimates for known and recovered chemical warfare materiel. The Acting Deputy further stated that the Product Manager was not assigned the mission of planning future remediation activities, nor is the mission included in the Chemical Agent Munitions Destruction appropriations funding level.

Audit Response. The actions taken by the Director, U.S. Army Chemical Materials Agency are responsive to the recommendation. We commend the Product Manager for Non-Stockpile Chemical Materiel for being proactive in planning for the remediation of buried chemical warfare materiel.

Appendix A. Scope and Methodology

In this third and final report addressing the Chemical Demilitarization Program (the Demilitarization Program) we evaluated reasons for the continuing cost growth within the Demilitarization Program and DoD plans and actions to dispose of buried chemical warfare materiel. As a result, we focused on factors affecting Demilitarization Program cost and schedule and DoD efforts to dispose of buried munitions. We performed this audit from January 2003 through May 2003 in accordance with generally accepted government auditing standards.

To evaluate whether the Office of the Secretary of Defense and the Army were effectively managing the Demilitarization Program, we examined "National Defense Authorization Act of 1993," (Public Law 102-484); section 8065, title 6, United States Code, "Defense Authorization Act for Fiscal Year 1997," (Public Law 104-208; section 1521, title 50, United States Code, "Destruction of Existing Stockpile of Lethal Chemical Agents and Munitions," (Public Law 99-145); the "Strom Thurmond National Defense Authorization Act for Fiscal Year 1999," (Public Law 105-261; and Army Regulation 200-1, "Environmental Protection and Enhancement," February 21, 1997.

We reviewed documentation dated from August 1994 through March 2003 that we obtained from the Demilitarization Program Office; disposal facilities located at Aberdeen, Maryland, Tooele, Utah, Anniston, Alabama, Umatilla, Oregon, Pine Bluff, Arkansas, and Newport, Indiana; future disposal facilities at Pueblo, Colorado, and Blue Grass, Kentucky; and from the Joint Munitions Command, Rock Island, Illinois.

To accomplish the audit objectives, we took the following steps:

- To determine the key factors affecting program cost and schedule, we reviewed GAO Report No. NSIAD 97-18, "Chemical Weapons and Materiel: Key Factors Affecting Disposal Costs and Schedule," February 10, 1997; GAO Report No. NSIAD 00-80, "Chemical Weapons Disposal: Improvements Needed in Program Accountability and Financial Management," May 8, 2000; the current approved program baseline for the Demilitarization Program; the Budget Estimate Submission for FYs 2004 and 2005; Program Budget Decision 204, "Chemical Agents and Munitions Destruction," December 9, 2002; and cost and schedule data. Additionally, we interviewed representatives from the Director, Chemical Materials Agency's staff including the Project Managers for Chemical Stockpile Disposal and Chemical Stockpile Emergency Preparedness, the Product Manager for Non-Stockpile Chemical Materiel, and the Program Manager for Assembled Chemical Weapons Assessment. Also, we met with representatives from the Army Cost and Economic Analysis Center and the Cost Analysis Improvement Group.

- To determine the efforts planned to dispose of buried munitions, we reviewed Army Regulation 200-1, "Environmental Protection and Enhancement," February 21, 1997, "Management Guidance for the Defense Environmental Restoration Program," September 2001, and the non-stockpile buried chemical warfare materiel cost estimate as provided in a rough-order-magnitude cost estimate that the Product Manager for Non-Stockpile

Chemical Materiel prepared. Additionally, we evaluated the "National Defense Authorization Act for Fiscal Year 1993," section 176, "Report on Destruction of Non-Stockpile Chemical Materiel," (Public Law 102-484) to determine the Army's plans for destroying chemical warfare materiel. We also interviewed representatives from offices of the Deputy Under Secretary of Defense (Installations and Environment) and the Army Corps of Engineers to determine their responsibilities and actions for the Defense Environmental Restoration Program, and the Product Manager for Non-Stockpile Chemical Materiel to identify those installations with chemical stockpile disposal facilities where remediation of potential burial sites with chemical warfare materiel could delay closure of the chemical stockpile disposal facilities.

General Accounting Office High-Risk Area. The General Accounting Office has identified several high-risk areas in the DoD. This report provides coverage of the DoD Weapons Systems Acquisition high-risk area.

Management Control Program Review

DoD Directive 5010.38, "Management Control (MC) Program," August 26, 1996, and DoD Instruction 5010.40, "Management Control (MC) Program Procedures," August 28, 1996, require DoD organizations to implement a comprehensive system of management controls that provides reasonable assurance that programs are operating as intended and to evaluate the adequacy of the controls.

Scope of the Review of the Management Control Program. In accordance with DoD policy, acquisition managers are to use program cost, schedule, and performance parameters as control objectives to implement the requirements of DoD Directive 5010.38. Accordingly, we limited our review to management controls directly related to areas of requirements, threat assessments, program assessments, cost estimating, earned value management, maintenance planning and oversight, design and operational failure reviews, contract management, and supply support planning for the five programs under the Demilitarization Program that the Under Secretary of Defense (AT&L) established. We reviewed management's self-evaluation applicable to those controls.

Adequacy of Management Controls. In the two previous reports, we identified material management control weaknesses for the Demilitarization Program as defined by DoD Instruction 5010.40. In Inspector General, DoD Report No. D-2003-015, "A Revised Acquisition Program Baseline and Threat Assessment for the Chemical Demilitarization Program," October 30, 2002, we identified material management control weaknesses in that controls were not in place to ensure that the program baseline agreement was revised when a significant baseline breach was reported and that a system threat assessment was not prepared for the program. Recommendations A. and B.2. of Report No. D-2003-015, if implemented, will enable the Under Secretary of Defense (AT&L) to use the baseline concept in managing future program cost and schedule and the depot site security managers to establish fully effective security plans. In Inspector General, DoD Report No. D-2003-088, "Acquisition of the Chemical Demilitarization Program," May 12, 2003, management controls were insufficient in conducting program cost reviews; obtaining accurate cost and schedule information; ensuring that the contractor at the Tooele Chemical Agent Disposal Facility performed timely preventive maintenance; ensuring that facility project managers conducted effective operational failure reviews;

and identifying initial and replenishment spares early in facility development, assigning national stock numbers to replenishment spares needed at multiple facilities, and establishing an obsolescence program to identify replenishment spares. Recommendations A.1., A.2., B.1., B.2., C.1., C.2., D., E.1., E.2., and E.3. of Report No. D-2003-088, if implemented, will improve the overall management of the Demilitarization Program and provide information needed by acquisition decision makers to make fully informed investment decisions. A copy of this report will be provided to the senior official responsible for management controls in the Office of the Under Secretary of Defense (AT&L) and the Department of the Army.

Adequacy of Management Self-Evaluation. The Director, Chemical Materials Agency performed annual reviews of the five programs under the Demilitarization Program that were established as assessable units to satisfy the management control requirement. The Director based his annual statement of assurance on statements from the project managers. However, in their self-evaluations, the project managers did not identify the specific management control weaknesses that the audit identified in the two audit reports because the self-evaluations did not review those areas as part of the assessable units.

Appendix B. Prior Coverage

General Accounting Office (GAO)

GAO Report No. 02-890, "Chemical Weapons: Lessons Learned Program Generally Effective but Could Be Improved and Expanded," September 10, 2002

GAO Report No. 01-850, "Chemical Weapons: FEMA and Army Must Be Proactive in Preparing States for Emergencies," August 13, 2001

GAO Report No. NSIAD 00-80, "Chemical Weapons Disposal: Improvements Needed in Program Accountability and Financial Management," May 8, 2000

GAO Report No. NSIAD 97-91, "Chemical Weapons Stockpile: Changes Needed in the Management of the Emergency Preparedness Program," June 11, 1997

GAO Report No. NSIAD 97-18, "Chemical Weapons and Materiel: Key Factors Affecting Disposal Costs and Schedule," February 10, 1997

Inspector General of the Department of Defense (IG DoD)

IG DoD Report No. D-2003-088, "Acquisition of the Chemical Demilitarization Program," May 12, 2003

IG DoD Report No. D-2003-068, "Army Response to Chemical Agent Incident at Tooele Chemical Agent Disposal Facility," March 28, 2003

IG DoD Report No. D-2003-015, "A Revised Acquisition Program Baseline and Threat Assessment for the Chemical Demilitarization Program," October 30, 2002

IG DoD Report No. 99-136, "Government-Furnished Equipment Year 2000 Issues for Army Chemical Demilitarization," April 16, 1999

IG DoD Report No. 99-081 "Tooele Chemical Agent Disposal Facility Preparation for Year 2000," February 9, 1999

IG DoD Report No. 99-060, "Johnston Atoll Chemical Agent Disposal System Preparation for Year 2000," December 24, 1998

IG DoD Report No 98-051, "Chemical Event at Tooele Chemical Agent Disposal Facility," January 20, 1998

Army Audit Agency

Army Audit Agency Report No. 01-131, "Financial Management of the Chemical Demilitarization Program," January 4, 2001

Army Audit Agency Report No. 01-001, "Matrix Support Requirements for the Chemical Demilitarization Program," October 2, 2000

Army Audit Agency Report No. 00-346, "Engineering Change Process for the Chemical Stockpile Disposal Project; Aberdeen Proving Ground, Maryland," August 14, 2000

Army Audit Agency Report No. 00-205, "Military Interdepartmental Purchase Requests; Program Manager for Chemical Demilitarization," March 27, 2000

Army Audit Agency Report No. 99-221, "Chemical Stockpile Emergency Preparedness Program; Aberdeen Proving Ground, Maryland," April 16, 1999

Army Audit Agency Report No. 99-155, "Chemical Agent Inventory Controls; Aberdeen Proving Ground, Maryland," February 17, 1999

Army Audit Agency Report No. 99-97, "Recycling Contaminated Metal; Rock Island Arsenal, Rock Island, Illinois," December 31, 1998

Army Audit Agency Report No. 99-26, "Lessons Learned - Chemical Stockpile Disposal Project; Aberdeen Proving Ground, Maryland," November 9, 1998

Army Audit Agency Report No. 97-190, "Non-Stockpile Chemical Material Project; Aberdeen Proving Ground, Maryland," May 12, 1997

Army Audit Agency Report No. 97-42, "Johnston Atoll Chemical Agent Disposal System," November 21, 1996

Appendix C. Management Roles Within the Chemical Demilitarization Program

This section describes the management roles of key Army officials within the Demilitarization Program. The DoD established the Demilitarization Program to support the Assistant Secretary of the Army for Installations and Environment in destroying all chemical warfare-related materiel while ensuring maximum protection of the public, personnel involved in the destruction effort, and the environment. In February 2003, the Army provisionally restructured the key offices that manage the Demilitarization Program and will define the specific duties for each office not later than October 2003. Key officials within the Demilitarization Program include:

Director, Chemical Materials Agency. Section 1521, title 50, United States Code, "Destruction of Existing Stockpile of Lethal Chemical Agents and Munitions," (Public Law 99-145) designates the Army as the lead agent for the complete destruction of the chemical weapons stockpile and related non-stockpile materiel. As a result, the Army designated the Program Manager for Chemical Demilitarization as the principal manager responsible for preparing and updating the overall planning and budgeting details necessary to execute the operation of destroying the chemical weapons. Under the restructured program, the Army reassigned those responsibilities to the Director, Chemical Materials Agency. For program acquisition related program issues, the Director reports to Assistant Secretary of the Army (Acquisition, Logistics, and Technology), who, in turn, reports to the Under Secretary of Defense (AT&L). For disposal facility operations and emergency preparedness related issues, the Director reports to the Commanding General, Army Materiel Command.

Program Manager for the Elimination of Chemical Weapons. Under the Demilitarization Program's restructured management, the Army established the Program Manager for the Elimination of Chemical Weapons to oversee the development and test and evaluation of the chemical disposal facilities. The Program Manager for the Elimination of Chemical Weapons reports to the Director, Chemical Materials Agency.

Deputy Director for Plant Operations, Office of the Director, Chemical Materials Agency. The Army established the Deputy Director for Plant Operations to manage the chemical warfare materiel storage, the day-to-day operations of the chemical disposal facilities once they become operational, and to oversee the execution of the emergency preparedness program. The Deputy for Plant Operations reports to the Director, Chemical Materials Agency.

Project Manager for Chemical Stockpile Disposal. The Project Manager for Chemical Stockpile Disposal is responsible for constructing and systemizing chemical disposal facilities for the stockpiles of unitary munitions with chemical agents at five locations. During those phases, the Project Manager reports to the Program Manager for the Elimination of Chemical Weapons. Upon successful construction and systemization of each chemical disposal facility, stockpile destruction responsibility for those sites transfers to the Deputy for Plant Operations, Office of the Director, Chemical Materials Agency. Accordingly, the Project Manager reports on those matters to the Deputy Director for Plant Operations.

Product Manager for Non-Stockpile Chemical Materiel. The Product Manager for Non-Stockpile Chemical Materiel is responsible for destroying all non-stockpile chemical materiel or chemical warfare materiel that is not part of the unitary stockpile. The non-stockpile chemical materiel mission includes binary chemical weapons, former chemical weapon production facilities, recovered chemical warfare materiel, and miscellaneous chemical warfare materiel. The Product Manager for Non-Stockpile Chemical Materiel reports to the Program Manager for the Elimination of Chemical Weapons.

Project Manager for Alternative Technologies and Approaches. Public Law 102-484 requires the Army to establish the position of Project Manager for Alternative Technologies and Approaches to examine alternative technologies for demilitarizing chemical weapons at two bulk facilities, if alternative operations can be completed within the baseline schedule, and if alternative operations are significantly safer and is equal to or more cost-effective than the approved baseline incineration process. The Project Manager for Alternative Technologies and Approaches reports to the Program Manager for the Elimination of Chemical Weapons.

Project Manager for Chemical Stockpile Emergency Preparedness. Public Law 99-145 requires the Chemical Stockpile Emergency Preparedness Project to ensure that the population surrounding the chemical storage facilities receives maximum protection. The Deputy for Plant Operations, Office of the Director, Chemical Materials Agency oversees the efforts of the Project Manager for Chemical Stockpile Emergency Preparedness. The "Strom Thurmond National Defense Authorization Act for Fiscal Year 1999" (Public Law 105-261), directs the Army to take responsibility for on-post emergency preparedness and the Federal Emergency Management Agency to take responsibility for off-post emergency preparedness. As a result, the Army established a memorandum of understanding with the Federal Emergency Management Agency to ensure that the Agency provided local municipalities with funding for the planned emergency preparedness.

Program Manager for Assembled Chemical Weapons Assessment. In 1996, in response to direction from Congress, the Office of the Secretary of Defense established the Program Manager for Assembled Chemical Weapons Assessment. The Congress directed that a program manager other than the Program Manager for Chemical Demilitarization identify and demonstrate at least two alternative technologies for destroying assembled chemical weapons. The Program Manager for Assembled Chemical Weapons Assessment has the responsibility for identifying and demonstrating the alternative technologies, and provides reports on program status directly to the Under Secretary of Defense (AT&L).

Appendix D. Environmental Restoration Programs for Chemical Warfare Materiel Burial Sites

The paragraphs that follow describe the Non-Stockpile Chemical Materiel Product and the DoD Environmental Restoration Programs.

Non-Stockpile Chemical Materiel

The Non-Stockpile Chemical Materiel Product is divided into five categories: binary chemical weapons, recovered chemical warfare materiel, buried chemical warfare materiel, former chemical weapons production facilities, and miscellaneous chemical warfare materiel. Most of the known chemical warfare materiel classified as non-stockpile, the U.S. has declared and is subject to the destruction requirements established under the provisions of the Chemical Weapons Convention. The warfare materiel that is not required for destruction under the Convention includes empty ton containers, recovered chemical agent identification sets, and buried chemical warfare materiel. Any time the Army recovers buried chemical warfare materiel, it must first be assessed to determine whether it meets the definition and criteria for a chemical weapon. If the recovered chemical warfare materiel meets the definition and criteria under the Convention, it will be declared and destroyed under the appropriate treaty verification regime.

Environmental Restoration Programs

Public Law 102-484 directs the Army to plan for the remediation of buried chemical warfare materiel sites that are a significant environmental concern to the public. To address this challenge, the DoD formally established the Defense Environmental Restoration Program in 1986 to provide for the cleanup of DoD hazardous waste sites. The DoD budgets about $3 billion annually to accomplish the program objectives. As with other DoD functions that require the handling of chemical warfare materiel, the Congress established the Army, as Executive Agent. The U.S. Army Environmental Center and the U.S. Army Corps of Engineers were tasked with overseeing the established environmental restoration programs that protect human health and the environment, clean up contaminated sites as quickly as resources permit, and to expedite cleanup to facilitate disposal of excess DoD properties for local reuse.

Service Installation Restoration Programs. Each of the Services has established Installation Restoration Programs to comply with the Defense Environmental Restoration Program. The mission of the Service programs is to identify, investigate, and clean up contamination at active and operating Service installations. Each installation commander has the overall responsibility for their sites, and coordinates the program execution, guidance, planning, oversight, and reporting of environmental cleanup with their respective environmental centers. The Army Environmental Center, the Naval Facilities Engineering Command, the

Air Force Center for Environmental Excellence, and the Defense Logistics Agency assist the installation commanders with planning for the remediation of sites where buried chemical warfare materiel has been identified. Within the Army, the U.S. Army Corps of Engineers is responsible for coordinating with the installation commanders, the excavation of the buried materiel from active sites.

The Product Manager for Non-Stockpile Chemical Materiel plans to perform program disposal tasks under service agreements with DoD activities and installations and expects reimbursement for the disposal of recovered chemical warfare materiel. The DoD has not identified the processes that will be used for reimbursement. Through April 2003, the Office of the Deputy Under Secretary of Defense (Installations and Environment) has not received tasking to prepare a comprehensive life-cycle cost estimate that identifies the requirements for the remediation of known or potential burial sites located on active Service installations.

Formerly Used Defense Sites Program. In September 2001, through the updated Defense Environmental Restoration Program Guidance, the Deputy Under Secretary of Defense (Installations and Environment) emphasized the need for the DoD to assess whether the formerly used defense sites contain buried chemical warfare materiel. As a part of that assessment, the DoD was to include an estimate of costs to complete recovery and disposal of the identified materiel. In January 2002, the Deputy Assistant Secretary of the Army (Environment Safety and Occupational Health) tasked the U.S. Army Corps of Engineers with initiating and completing a plan to accelerate the schedule for the recovery of chemical warfare materiel at the sites by the end of FY 2004. The Corps of Engineers assigned this task to its Huntsville Engineering and Support Center and directed the center to provide periodic reporting on the plan's implementation. The Product Manager for Non-Stockpile Chemical Materiel is responsible for the storage, transportation, and destruction of the chemical warfare materiel once it is recovered from the sites.

Base Realignment and Closure Program. The Base Realignment and Closure Program is responsible for environmental restoration at all installations closed under the Base Realignment and Closure Act. Army Regulation 200-1, "Environmental Protection and Enhancement," February 21, 1997, assigns the U.S. Army Environmental Center as program manager for the remediation of chemical warfare materiel at all closure sites. Further, the Regulation states that the U.S. Army Corps of Engineers is responsible for the excavation of the chemical warfare materiel. The Product Manager for Non-Stockpile Chemical Materiel is responsible for disposal of the recovered chemical warfare materiel.

Use of Chemical Stockpile Disposal Facilities. The Assistant Secretary of the Army (Installations and Environment) recently expressed an interest in four chemical warfare materiel burial sites at installations scheduled for closure because of their co-location with existing or planned chemical stockpile disposal facilities. The four sites are the Newport Chemical Depot, Indiana; Umatilla Chemical Depot, Oregon; Pueblo Chemical Depot, Colorado; and, the Deseret Chemical Depot, Utah. The Product Manager for Non-Stockpile Chemical Materiel is evaluating the possibility of using the chemical stockpile disposal facilities for destruction of the chemical warfare materiel recovered from the four burial sites.

Appendix E. Report Distribution

Office of the Secretary of Defense

Under Secretary of Defense for Acquisition, Technology, and Logistics
  Assistant to the Secretary of Defense (Nuclear and Chemical and Biological Defense Programs)
Under Secretary of Defense (Comptroller)/Chief Financial Officer
  Deputy Comptroller (Program/Budget)
  Director (Program Analysis and Evaluation)

Joint Staff

Director, Joint Staff

Department of the Army

Assistant Secretary of the Army (Acquisition, Logistics, and Technology)
  Director, Chemical Materials Agency
    Program Manager, Elimination of Chemical Weapons
Assistant Secretary of the Army (Installations and Environment)
Commander, Army Materiel Command
  Commander, U.S. Army Soldier Biological and Chemical Command
Auditor General, Department of the Army
Commander, Army Test and Evaluation Center

Department of the Navy

Auditor General, Department of the Navy

Department of the Air Force

Auditor General, Department of the Air Force

Other Defense Organization

Director, Defense Contract Management Agency

Non-Defense Federal Organization

Office of Management and Budget

Congressional Committees and Subcommittees, Chairman and Ranking Minority Member

Senate Committee on Appropriations
Senate Subcommittee on Defense, Committee on Appropriations
Senate Committee on Armed Services
Senate Committee on Governmental Affairs
House Committee on Appropriations
House Subcommittee on Defense, Committee on Appropriations
House Committee on Armed Services
House Committee on Government Reform
House Subcommittee on Government Efficiency and Financial Management, Committee on Government Reform
House Subcommittee on National Security, Emerging Threats, and International Relations, Committee on Government Reform
House Subcommittee on Technology, Information Policy, Intergovernmental Relations, and the Census, Committee on Government Reform

Department of the Army Comments

DEPARTMENT OF THE ARMY
OFFICE OF THE ASSISTANT SECRETARY OF THE ARMY
ACQUISITION LOGISTICS AND TECHNOLOGY
103 ARMY PENTAGON
WASHINGTON DC 20310-0103

AUG - 6 2003

SAAL-ZC

MEMORANDUM FOR DEPARTMENT OF DEFENSE INSPECTION GENERAL, OFFICE OF THE (ACQUISITION MANAGEMENT DIRECORATE)

SUBJECT: DODIG Report No. D-2003-0070, The Chemical Demilitarization Program: Increased Costs for Stockpile and Non-stockpile Chemical Materiel Disposal Programs.

Management comments from the Acting Deputy Assistant Secretary of the Army for the Elimination of Chemical Weapons (Enclosure 1) and the Director, U. S. Army Chemical Materials Agency (Enclosure 2) for the subject report are enclosed.

Nancy M. Ray
Colonel, U.S. Army
Acting Deputy Assistant Secretary of the Army
for the Elimination of Chemical Weapons

Enclosures

DoDIG DRAFT OF A PROPOSED REPORT – DATED MAY 30, 2003
D2003AE-0070

The Chemical Demilitarization Program: Increased Costs for Stockpile and Non-Stockpile Chemical Materiel Disposal Programs

U.S. Army Chemical Materials Agency (Provisional)
Comments to the Recommendations and Other Comments

RECOMMENDATION B.

B.2. *We recommend that the Product Manager for Non-Stockpile Chemical Materiel update the plan and cost estimate for disposal of buried munitions after the environmental offices of the DoD components implement Recommendation B1.*

July 28, 2003 Reply to the Proposed Draft Report:

- Concur. The Product Manager for Non-Stockpile Chemical Materiel (PM NSCM) is currently doing estimates for four (4) potential BRAC locations[1], and is prepared to update the estimates for all burial sites. During the fourth quarter of fiscal year (FY) 2003, PM NSCM will meet with representatives from the office of the Under Secretary of Defense for Acquisition, Technology, and Logistics to discuss burials and the path forward.

Other Comments:

July 28, 2003 Reply to the Proposed Draft Report:

- We recommend that DoDIG reexamine statements made on pages 3 and 9 of the draft report relative to closure costs. Cost estimates for the closure of incineration-based disposal facilities are not expected to negatively affect future program cost estimates. The closure cost estimates used to revise the life cycle cost estimate in September 2001 were based on an estimated 33-month duration for JACADS Closure at a cost of $411 million. With JACADS Closure now expected to be completed 2 1/2 months early at a cost of approximately $365 million, closure estimates submitted in September 2001 for other incineration-based stockpile disposal facilities are expected to be adequate based on JACADS lessons learned.

- On page 7, the draft report discusses additional funding for emergency preparedness requested by the State of Alabama. It should be noted the Army has funded the additional FY 2003 collective protection and special population requirements identified by the State.

[1] The four locations are Deseret Chemical Depot, UT; Newport Chemical Depot, IN; Pueblo Chemical Depot, CO; and Umatilla Chemical Depot, OR.

Encl 1

DoDIG DRAFT OF A PROPOSED REPORT – DATED MAY 30, 2003
D2003AE-0070

The Chemical Demilitarization Program: Increased Costs for Stockpile and Non-Stockpile Chemical Materiel Disposal Programs

U.S. Army Chemical Materials Agency (Provisional)
Comments to the Recommendations and Other Comments

Other Comments: (continued)

July 28, 2003 Reply to the Proposed Draft Report:

- Clarification is provided relative to cost estimates for disposal of buried chemical munitions. The $11.7 billion cost estimate for disposal of buried munitions cited on page 12, and to which the footnote 6 refers, has not been updated, except for changes to inflation indices, since 1996. New cost estimates utilizing current remediation and treatment procedures, technology, site information and environmental standards will significantly increase the cost estimate.

DoDIG DRAFT OF A PROPOSED REPORT – DATED MAY 30, 2003
D2003AE-0070

The Chemical Demilitarization Program: Increased Costs for Stockpile and Non-Stockpile Chemical Materiel Disposal Programs

Acting DASA(ECW)
Comments to the Recommendations

RECOMMENDATION B2.

B.2. *We recommend that the Product Manager for Non-Stockpile Chemical Materiel update the plan and cost estimate for disposal of buried munitions after the environmental offices of the DoD components implement Recommendation B1.*

Reply to the Proposed Draft Report:

- Non-concur with the recommendation as written. The Product Manager for Non-Stockpile Chemical Materiel (PM NSCM) should participate as subject matter experts in updating the cost estimates for known, recovered materiel. PM NSCM does not currently have the assigned mission of planning future remediation activities, nor is that mission included in the Chemical Agent Munitions Destruction appropriations funding level.

Encl 2

Team Members

The Acquisition Management Directorate, Office of the Assistant Inspector General for Auditing of the Department of Defense prepared this report. Personnel in the Office of the Inspector General of the Department of Defense who contributed to the report are listed below.

Mary L. Ugone
John E. Meling
Rodney D. Britt
Melinda M. Oleksa
Chrispian M. Brake
Shaun B. Jeffery
Jacqueline N. Pugh

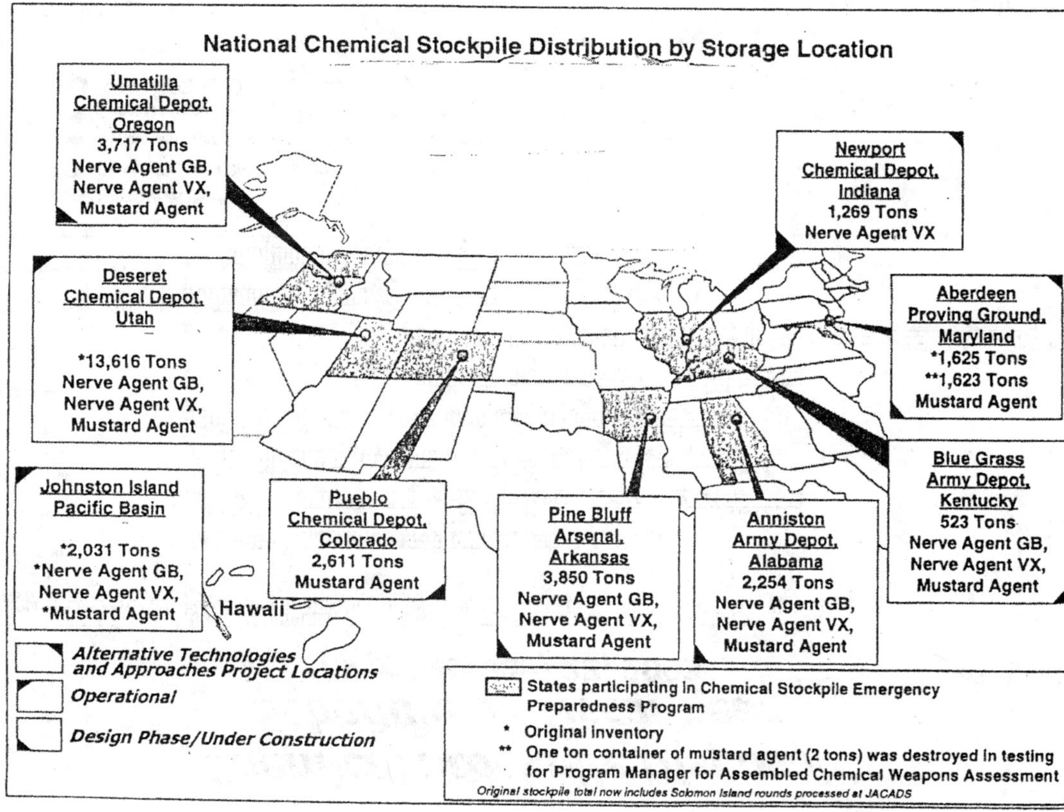
National Chemical Stockpile Distribution by Storage Location
Umatilla Chemical Depot, Oregon
3,717 Tons
Nerve Agent GB, Nerve Agent VX, Mustard Agent
Deseret Chemical Depot, Utah
*13,616 Tons
Nerve Agent GB, Nerve Agent VX, Mustard Agent
Johnston Island Pacific Basin
*2,031 Tons
*Nerve Agent GB, Nerve Agent VX, *Mustard Agent
Hawaii
Pueblo Chemical Depot, Colorado
2,611 Tons
Mustard Agent
Pine Bluff Arsenal, Arkansas
3,850 Tons
Nerve Agent GB, Nerve Agent VX, Mustard Agent
Anniston Army Depot, Alabama
2,254 Tons
Nerve Agent GB, Nerve Agent VX, Mustard Agent
Newport Chemical Depot, Indiana
1,269 Tons
Nerve Agent VX
Aberdeen Proving Ground, Maryland
*1,625 Tons
**1,623 Tons
Mustard Agent
Blue Grass Army Depot, Kentucky
523 Tons
Nerve Agent GB, Nerve Agent VX, Mustard Agent
Alternative Technologies and Approaches Project Locations
Operational
Design Phase/Under Construction
States participating in Chemical Stockpile Emergency Preparedness Program
* Original inventory
** One ton container of mustard agent (2 tons) was destroyed in testing for Program Manager for Assembled Chemical Weapons Assessment
Original stockpile total now includes Solomon Island rounds processed at JACADS

Chemical Stockpile Disposal Schedule by Fiscal Year

As of February 23, 2003

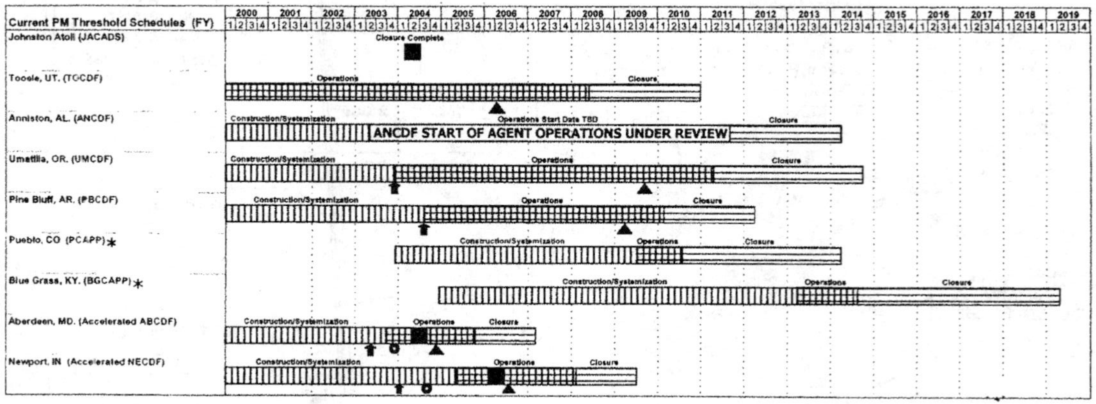

- * -Schedule as Provided by PM ACWA
- ↑ -Denotes Program Office Estimate Start of Operations
- ▲ -Denotes Program Office Estimate Completion of Operations
- ◆ -Denotes Current PM Threshold Agent Neutralized
- O -Denotes Program Office Estimate Agent Neutralized

Pres2348 - 11 - 3/21/2003 8:24:47 AM

NSCMP Milestone Schedule

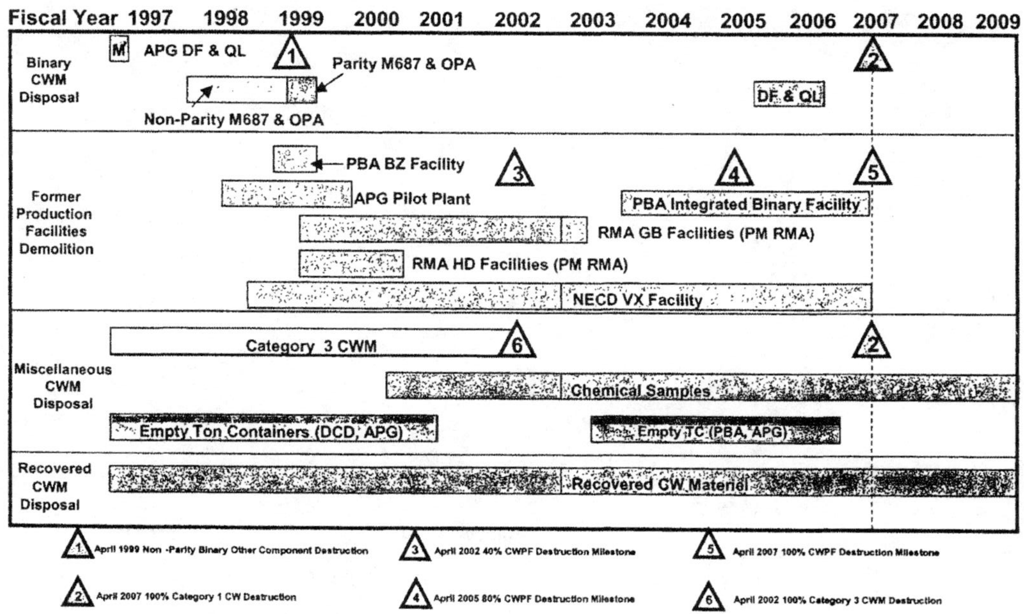

Pres2348 - 12 - 3/21/2003 8:24:47 AM

QUESTIONS AND ANSWERS SUBMITTED FOR THE RECORD

OCTOBER 30, 2003

QUESTIONS SUBMITTED BY MR. SAXTON

Mr. SAXTON. As noted in both Mr. Hinton's and Mr. Wakefield's statements, the Assembled Chemical Weapons Assessment (ACWA) Program is managed separately from the chemical stockpile destruction program, as directed by Congress, and maintenance of a separete program contributes to a complex program management structure. In his statement Mr. Wakefield said that the Department of Defense is looking at further streamlining the management of the chemical demilitarization program and now desires to consolidate the ACWA program manager under the Army's Chemical Materials Agency. How does the current division of the program structure between the Chemical Weapons Alternative adversely affect the program and how would consolidation of these two activities as proposed by Mr. Wakefield improve the overall management of the program? What legislative action would be required to make such a change? [1]

Mr. HINTON. Our previous work has shown that the division in the program's structure between the ACWA program and the chemical stockpile destruction program has contributed to ineffective coordination and communication, inefficient efforts, and obscured accountability. For example, in 2000, we reported that the chemical demilitarization program had a complex stucture that seperates management roles, resposibilities, and accountability for achieving program results that contributed to ineffective coordination and communication between ACWA and the rest of the program.[2] In 2002, we also reported that there were barriers to communicating lessons learned between ACWA and the rest of the program, resuting in duplication of efforts.[3] As long as this separation exists between ACWA and the rest of the program, the management structure will remain complex and difficult to manage.

While our previous work has not commented on whether the ACWA program should be consolidated with the rest of the program, we believe that consolidating these two programs could result in some improvements in program management, provided that the consolidated program also develops and implements an overarching strategy and implementation plan. While consolidation should provide a number of benefits—such as simplifying the management structure, reducing duplication of efforts, and improving coordination and communication—these benefits will be limited if the program does not successfully develop a roadmap for success.

Finally, as acknowledged by DOD in its statement to the Committee, consolidating the two programs requires amending existing legislation mandating independent management of the ACWA program and directing the program manager for ACWA to manage the chemical demilitarization activities at Blue Grass and Pueblo Depot. For example, the Department of Defence Appropriations Act, 2003 (Pub. L. No. 107-248 (2002)), which designates the program manager for ACWA to be the manager for Blue Grass and Pueblo activities, and earlier ACWA legislation must be amended to reflect the program management changes DOD seeks.

Mr. WAKEFIELD. As noted by the General Accounting Office in its September 2003 audit report, CHEMICAL WEAPONS: Sustained Leadership, Along With Key Strategic Management Tools, Is Needed to Guide DOD's Destruction Program, "ACWA, as congressionally directed, will continue to be managed separately from the Army DOD". This division of program management structure adversely affects the program by dividing accountability throughout multiple organizations and creating duplicative functions that unnecessarily inflate manpower and resource requirements.

Consolidation of the Assembled Chemical Weapons Alternatives Program under the Chemical Materials Agency would significantly improve the overall management of the Chemical Demilitarization Program by consolidation accountability, simplify-

[1] U.S. General Accounting Office, *Chemical Weapons: Better Management Tools Needed to Guide DOD's Stockpile Destruction Program,* GOA-04-221T (Washington, D.C.: October 30, 2003).

[2] U.S. General Accounting Office, *Chemical Weapons Disposal: Improvements Needed in Program Accountability and Finacial Management,* GAO/NSAID-00-80 (Washington, D.C.: May 8, 2000).

[3] U.S. General Accounting Office, *Chemical Weapons: Lessons Learned Program Generally Effective but Could Be Improved and Expanded,* GAO-02-890 (Washington, D.C.: September 10, 2002).

ing the chain of command, restructuring and combining organizational functions to reduce redundancies, and making the executive agent of the program, the Army, responsible for the program in its entirety.

The Department recommends Congress incorporate any proposed legislative language, with regard to consolidation of the Assembled Chemical Weapons Alternatives Program, into the Dedense Authorization Bill. Draft language is currently under review by the Office of Management and Budget.

Secretary BOLTON. The Army defers to the on going discussions between the OSD and Congressional staff and stands ready to support a Congressional initiative to merge the ACWA program into the Army's program under the Chemical Materials Agency.

Again, the Army defers to the on going discusions between the OSD and Congressional staff regarding this matter. The legislative basis for the current program structure is section 142 of Public Law 105-261 and section 101(b) of Public Law 104-208. Both codified at 50 U.S.C. 1521 note.

Mr. PARKER. As the Program Manager for the ACWA Program, I believe this consolidation would be beneficial in a businedd context. Merging of ACWA into the Army program under CMA would offer the potential for business efficiency by bringing together some of the common overhead type functions and reducing the staff accordingly. Any actions of this type would be subject to changing of the statute.

There would be a requirement to change the enabling statute for the ACWA program.

Mr. SAXTON. In your statement you spoke briefly of the application of lesson learned at one chemical demilitarization and storage site to other sites. How are the lessons learned from the experiences gained in the operations and closure of the Johnston Atoll pilot plant and the operations of the 2nd-generation Tooele facility being applied to the design and operation of the 3rd-generation incineration facilities at Anniston, Umatilla, and Pine Bluff?

Are any of the lessons learned from the design and operation of the first and second generation facilities applicable to those chemical weapons destruction facilities that will use alternative technologies?

Mr. WAKEFIELD. A significant number of key lessons learned have been transferred from the Johnston Atoll pilot plant to the second generation Tooele Facility and from both to the third generation facilities of Anniston, Umatilla, and Pine Bluff. The lessons learned are not limited to equipment modifications and improvements, but include improvements in contracting approaches and partnership with the systems contractors. These lessons learned have not only resulted in improvements in plant efficiencies and destruction performance, but more importantly, have resulted in a workforce culture shift at each facility that has dramatically improved safety performance.

Many lessons learned from the design and operation of the incineration facilities are applicable to destruction facilities that are or will be using alternative technologies. For example, the Newport and Aberdeen alternative technology sites have benefited from lessons learned in ventilation, monitoring, contamination control, and other technology neutral areas.

The Systems Contractors for Pueblo and Blue Grass also benefit from the lessons learned program because many of the lessons being transferred are not technology dependent. They are using many of the design concepts, construction designs and techniques, and startup approaches developed for the baseline facilities

Mr. PARKER. A significant number of key lessons learned have been transferred from the Johnston Atoll pilot plant to the second generation Tooele Facility and from both to the third generation facilities of Anniston, Umatilla, and Pine Bluff. The lessons learned are not limited to equipment modifications and improvements, but include improvements in contracting approaches and partnership with the systems contractors. These lessons learned have not only resulted in improvements in plant efficiencies and destruction performance, but more importantly, have resulted in a workforce culture shift at each facility that has dramatically improved safety performance.

Lessons from earlier generation facilities are shared in a variety of ways. Later generation facilities have benefited from knowledge of the many equipment and facility improvements made by earlier generation facilities, and many of these lessons have been incorporated into the design packages of later generation facilities.

A key aspect of the existing lessons learned program is to encourage communications among all operating contractors, particularly among the engineers, technicians, operators and maintenance personnel responsible for facility operations. The recent very successful start up and initial operation of the Anniston facility is objective evidence of the success of the lessons learned program. Anniston had a flawless startup and has processed significantly more.

Monthly meetings allow subject matter experts to discuss common problems and to develop site-specific solutions. Meetings to discuss common problems and seek solutions are also conducted at the management level. These senior level meetings are focused on building contractor ownership and a top down lessons learned process. The intent is to ensure that lessons learned are effectively implemented at the plant by the systems contractors who execute disposal operations.

Many lessons learned from the design and operation of the incineration facilities are applicable to destruction facilities that are or will be using alternative technologies. For example, the Newport and Aberdeen alternative technology sites have benefited from lessons learned in ventilation, monitoring, contamination control, and other technology neutral areas.

The systems Contractors are encouraged to work together with program staff to identify and manage risk. The systems contractors for Pueblo and Blue Grass also benefit from the lessons learned program because many of the lessons being transferred are not technology dependent. They are using many of the design concepts, construction designs and techniques, and startup approaches developed for the baseline facilities. In addition, these Systems contractors are actively involved in CMA's lessons learned program and participate in the monthly subject matter expert discussions.

Mr. SAXTON. Among the factors affecting cost and schedule growth in the chemical demilitarization program, the DOD Inspector General cites costly delays in reaching public consensus with obtaining state permit modifications needed for beginning disposal operations, the role of state and local governments, and the role of special interest groups. In your prepared statements, you cite improvements in management and coordination of the Chemical Stockpile Emergency Preparedness Program (CSEPP) between the Army and the Federal Emergency Management Agency and how these improvements in the program have helped state and local communities become better prepared to respond to chemical emergencies. Despite these gains, however, the GAO states that CSEPP Program costs are rising because some states have expanded their preparedness requests beyond the approved budgets. The GAO's testimony indicated that the most recent data shows unfunded requirements in fiscal years 2004 and 2005 of $39.4 million and $49.0 million, respectively.

What is the significance to the chemical demilitarization program of projected unfunded requirements in the CSEPP program? What is the plan for meeting unfunded but validated CSEPP requirements during the budget year? The execution year?

What further improvements are being considered in management and administration of the CSEPP program?

What measures are being taken to ensure that state and local requests meet the objectives and requirements of the program?

Mr. WAKEFIELD. In the past, the chemical demilitarization program has had to absorb FEMA-validated unfunded CSEPP requirements. During both budget and execution years, FEMA-validated requirements that were not anticipated in the budget process are funded from the Chemical Agent & Munitions Destruction account. This reprogramming of funds directly affects the chemical demilitarization schedule, resulting in program slippage and higher costs for both incineration and alternative technologies sites and for the CSEPP in the long run.

A key element in these areas is early identification of expected costs. The Army and FEMA have and will continue to stress to the off-post communities the need to identify their budget requirements early in the budget cycle. In addition, FEMA is working closely with off-post communities to help them refine their budgeting processes. The Army and FEMA have built a strong partnership to assist CSEPP communities and, over the years, has made the pursuit of new ideas in management improvements a program mainstay.

The Army requires that all state, local and tribal requests be validated by FEMA as meeting program benchmarks for preparedness. FEMA reviews benchmark status on a quarterly basis.

Mr. PARKER. In the past, the chemical demilitarization program has had to absorb FEMA-validated unfunded CSEPP requirements. During both budget and execution years, FEMA-validated requirements that were not anticipated in the budget process are funded from the Chemical Agent & Munitions Destruction, Army (CAMD, A) account. This reprogramming of funds directly affects the chemical demilitarization schedule, resulting in program slippage and higher costs for both incineration and alternative technologies sites and for the CSEPP in the long run.

A key element in these areas is early identification of expected costs. The Army and FEMA have and will continue to stress to the off-post communities the need to identify their budget requirements early in the budget cycle. In addition, FEMA is working closely with off-post communities to help them refine their budgeting processes. The Army and FEMA have built a strong partnership to assist CSEPP communities and, over the years, has made the pursuit of new ideas in management improvements a program mainstay.

The Army requires that all state, local and tribal requests be validated by FEMA as meeting program benchmarks for preparedness. FEMA reviews benchmark status on a quarterly basis.

Mr. SAXTON. As a result of the terrorist attack on the United States on September 11, 2001, a number of measures were taken to increase the security of the chemical stockpile storage and demilitarization sites, including increasing the number of Army civilian security guards at each installation and assigning National Guard soldiers to chemical site defense force and installation defense force missions.

What are the program and budget implications to the chemical demilitarization program and to the Department of the Army of the establishment of the increased security forces, both DA civilian security guards and Army National Guard soldiers?

Mr. WAKEFIELD Since September 11, 2001, CMA has not identified any direct program or budget impacts to the CAMD, A appropriation (Chemical Demilitarization Program) caused by the establishment of heightened security postures at each of our chemical weapons storage sites. The Operation and Maintenance, Army (O&M, A) appropriation that finances chemical stockpile storage, however, has been significantly impacted. This impact is focused on the increased. logistical and manpower requirements associated with augmented security measures identified for the stockpile storage sites. A subsequent reimbursement to the O&M, A was contained in a Supplemental appropriation provided to finance overall costs of the Global War On Terrorism (GWOT).

The need for increased security at the storage and demilitarization sites will continue and will present significant future requirements to be addressed in the department's programming and budgeting process. We estimate the FY05 cost for maintaining increased security postures at our storage sites at $23M, to be paid out of the military pay appropriation. On-going efforts to reduce the stress on the National Guard forces that currently perform the stockpile security will likely result in transferring the mission to DOD or contract security. The associated cost to the storage program may increase to as much as $75M, to be paid out of a different account.

Mr. PARKER. Since September 11, 2001, CMA has not identified any direct program or budget impacts to the CAMD, A appropriation (Chemical Demilitarization Program) caused by the establishment of heightened security postures at each of our chemical weapons storages sites. The Operation and Maintenance, Army (O&M, A) appropriations that finances chemical stockpile storage, however, has been significantly impacted. This impact is focused on the increased logistical and manpower requirements associated with augmented security measures identified for the stockpile storage sites. A subsequent reimbursement to the O&M, A was contained in a Supplemental appropriation provided to finance overall costs of the Global War On Terrorism (GWOT). The need for increased security at the storage and demilitarization sites will continue and will present significant future requirements to be addressed in the department's programming and budgeting process. We estimate the FY05 cost for maintaining increased security postures at our storage sites at $23M, to be paid out of the military pay appropriation. On-going efforts to reduce the stress on the National Guard forces that currently perform the stockpile security will likely result in transferring the mission to DOD or contract security. The associated cost to the storage program may increase to as much as $75M, to be paid out of a different account.

Mr. SAXTON. What actions are being taken by the Department of the Army to evaluate new equipment and methods for improving both near-real time monitoring and long term monitoring and improve technologies and equipment to increase specificity lower detection limits and response times, and minimize false alarms?

How are these activities being coordinated within the chemical agent detection community?

Mr. WAKEFIELD. The Army has a program to evaluate emerging monitoring technologies for battlefield and demilitarization applications. The CMA, in conjunction with the Edgewood Chemical and Biological Center (ECBC), looks at these technologies to ascertain their effectiveness as compared to fielded systems. When a new technology is identified, the CMA specifically evaluates it for demilitarization and storage applications. If the technology demonstrates greater fidelity and capability than an existing monitoring system, an acquisition strategy is developed to procure and field the required equipment. To date, no system has been identified that successfully and consistently outperforms the existing monitors in place at our nation's storage and demilitarization sites.

The Army works hand in hand with the Centers for Disease Control and Prevention and the National Research Council to evaluate and test our existing systems as compared to what is available in the commercial sector. Industry is involved through the testing conducted by ECBC and the CMA EMO—as new processes and systems are developed, they are tested and evaluated. Those systems that are identified as promising undergo additional testing to determine if they exceed the capabilities of our fielded systems. This information is shared throughout industry and within the Departments of Defense and of the Army.

Secretary BOLTON. The Army has a program to evaluate emerging monitoring technologies for battlefield and demilitarization applications. The CMA, in conjunction with the Edgewood Chemical and Biological Center (ECBC), looks at these technologies to ascertain their effectiveness as compared to fielded systems. When a new technology is identified, the CMA specifically evaluates it for demilitarization and storage applications. If the technology demonstrates greater fidelity and capability that an existing monitoring system, an acquisition stratagy is developed to procure and field the required equipment. To date, no system has been identified that successfully and consistently outperforms the existing monitors on place at our nation's storage and demilitarization sites.

The Army works hand in hand with the Centers for Disease Control and Prevention and the National Research Council to evaluate and test our existing systems as compared to what is available in the commercial sector. Industry is involved through the testing conducted by ECBC and the CMA EMO—as new proceses and systems are developed, they are tested and evaluated. Those systems that are identified as promising undergo additional testing to determine if they exceed the capabilities of our fielded systems. This information is shareed throughout industry and within the Departments of Defense and of the Army.

Mr. PARKER. The data produced by the current monitoring systems used at all chemical stockpile and storage sites ensure that all agent operations are fully protective of human health and the environment and meet all applicable standards. The Department of Health and Human Services, National Research Council, and the individual states have reiterated this fact through statements and operational permit inspections. Within the Department of the Army, the CMA's Environmental and Monitoring Office (EMO) has an active program that investigates new technologies as they become available that may lend themselves to improved chemical agent detection in our facilities. The technology used to currently monitor at the facilities was developed decades ago and is still the state of the art for detecting very low levels of chemical warfare agents. Our program has improved the technology and equipment over time to lower the detection limits and minimize interferences (false alarms). There is no different technology commercially available that would be able to monitor at the proposed Department of Health and Human Services (DHHS) Airborne Exposure Levels prescribed for our programs in the 10-15 second time frame. In addition, EMO has two multimillion dollar efforts ongoing to evaluate new Commercial Off the Shelf (COTS) equipment and methods for improving both near-real time monitoring and historical long teem monitoring. The COTS equipment being evaluated still uses the same basic technology that is currently used. Fielding of equipment in thefuture and method changes will be based on the results from these tasks along with operational and funding constraints.

The current false alarm rate for near real-time (NRT) instruments is already extremely low. The likelihood of a false alarm at any of the demilitarization sites is approximately 1 in 20,000 instrument analytical cycles. This is a successful instrument operational performance of greater than 99.99%. As indicated by this number, the Army has worked extremely hard to minimize false alarms and will continue to strive for the lowest false alarm rate possible forany monitoring unit employed at the demilitarization site.

Equipment evaluations are reported regularly to the chemical agent detection community at various workshops, lessons learned sessions, monitoring roundtable discussions, environmental conferences, and reports circulated within the CMA. These information vehicles have served the demilitarization program well in the past and will continue to be used to disseminate chemical agent monitoring evaluation information.

In addition, CMA will host a technology workshop in January 2004 to explore with the commercial and academic communities any new or rapidly developing technology that would improve performance of our monitoring network. Workshops such as this are routinely held and widely advertised in aneffort to continually advance the state of the art in monitoring.

Mr. SAXTON. The National Defense Appropriation Act for Fiscal Year 2004 includes $1.5 billion for the chemical weapons and munitions destruction program and the Military Construction Appropriations Bill contains an additional $120 million for military construction at various stockpile storage and demilitarization sites. The appropriation represents a reduction of $30 million to the budget request for chemical demilitarization operations and maintenance and earmarks an additional $13 million of the budget request for specific elements within the program. How does this shortfall of approximately $43 million affect the Army's ability to execute the Fiscal Year 2004 chemical demilitarization program and what are the implications of this shortfall for the Fiscal Year 2005 program?

Mr. WAKEFIELD. Operations for fiscal year 2004 will not be severely impacted by the funding reduction because of carryover funds that resulted from schedule slips during fiscal year 2003. However, we do expect impacts upon FY 2005 activities at Aberdeen.

Within the $30 million reduction, $20 million was designated to be taken against the Aberdeen Chemical Demilitarization project. The project, however, had sufficient fiscal year 2003 carryover funds as a result of a modification to the original Aberdeen contract prior to the acceleration initiative. The $20 million reduction, thus, has not affected the Aberdeen site's ability to execute its FY 2004 program, which includes completing agent neutralization operations and beginning operations at the ton container cleanout facility.

With respect to the $10 million earmark specified in the FY 2004 Appropriations Act under the CAMD appropriation, funds were transferred from the Chemical Stockpile Disposal Project to the Chemical Stockpile Emergency Preparedness Program for execution in accordance with the Congressional language. The $10 million was available from the Chemical Stockpile Disposal Project as a result of cost avoidance realized by the early completion of closure activities at the Johnston Atoll Chemical Agent Disposal System.

Secretary BOLTON. Operations for fiscal year 2004 will not be severely impacted by the funding reduction because of carryover fund that resulted from schedule slips during fiscal year 2003. As detailed by Mr. Parker, however, we do expect impacts upon FY 2005 activities at Aberdeen.

Mr. Parker. Within the $30 million reduction $20 million was designated to be taken against the Aberdeen Chemical Demilitarization project. The project, however, had sufficient fiscal year 2003 carryover funds as a result of a modification to the original Aberdeen contract prior to the acceleration initiative. The $20 million reduction, thus, has not affected the Aberdeen site's ability to execute its FY 2004 program, which includes completing agent neutralization operations and beginning operations at the ton container cleanout facility.

Because it is anticipated that all budgeted FY 2005 funds will now be applied toward agent neutralization operations, the reduction may affect the FY 2005 program at Aberdeen, that includes the completion of ton container cleanout and commencement of closure activities.

With respect to the $10 million earmark specified in the FY 2004 Appropriations Act under the CAMD appropriation, funds were transferred from the Chemical Stockpile Disposal Project to the Chemical Stockpile Emergency Preparedness Program for execution in accordance with the Congressional language. The $10 million was available from the Chemical Stockpile Disposal Project as a result of cost avoidance realized by the early completion of closure activities at the Johnston Atoll Chemical Agent Disposal System.

In general terms, Congressional reduction to the President's Budget Request also reduces overall program flexibility in areas such as risk reduction and mitigation efforts, an area recently reported by the GAO to be deficient In addition, it limits the Army's ability to respond to technological advancement opportunities in such areas as monitoring.

Mr. SAXTON. When changes in program priorities and schedules cause changes in funding requirements within a budget year or in future program years, how are such changes accommodated within the program? Or, differently stated, what is the impact of unfunded requirements on the program and how are such requirements addressed?

Mr. WAKEFIELD. While the DOD Chemical Demilitarization Program is a Defense-Wide program, the program is managed and executed by the Army, who is designated the executive agent through Public Law 99-145. While exceptions have been made in the past, agencies responsible for the execution of DOD programs are generally responsible for rectifying the shortfalls in their programs. Occasionally the Department has some leeway to provide assistance with respect to DOD programs. However for the Chemical Demilitarization Program in Fiscal Year 2004, this has not been the case.

Within the Army, unfunded requirements are prioritized against existing program requirements in the Annual Funding Plan (AFP). If an unfunded requirement is determined to have a higher priority than existing program items, it is funded within the existing resources and lower priority activities are deferred, potentially resulting in schedule delays and increased life cycle costs. When funding an unfunded requirement within the existing AFP would result in a program breach (cost or schedule), the matter is presented to the Defense Acquisition Executive, who directs that the requirement either be funded within the current AFP, or directs the Army to program the funds for the following fiscal year. An unfunded requirement that does not have a higher priority than existing requirements would be addressed in the next program and budget cycle.

Mr. SAXTON. The fiscal year 2004 budget request included funding for the program in a new Appropriation Title "054 Other Defense Programs" that is separate from that of the Army or any of the other military departments, thereby meeting the guidance in section 1412 of the Department of Defense Authorization Act, 1986 (Public Law 99-145) that this program shall not be included in the budget accounts of any of the military departments. Funding the destruction program in a defense-wide account ensures that the program is subject to the appropriate level of management and oversight and ensures that the program is not subject to the internal budget priorities on one particular service.

When changes in the program priorities and schedules cause changes in funding requirements within a budget year or in future program years, how are such changes accommodated within the program?

Secretary BOLTON. Unfunded requirements are prioritized against existing program requirements in the Annual Funding Plan (AFP). If an unfunded requirement is determined to have a higher priority than existing program items, it is funded within the existing resources and lower priority activities are deferred, potentially resulting in schedule delays and increased life cycle costs. When funding an unfunded requirement within the existing AFP would result in a program breach (cost or schedule), the matter is presented to the Defense Acquisition Executive, who directs that the requirement either be funded within the current AFP or considered in the Program Objectives Memorandum (POM) for the following fiscal year. An unfunded requirement that does not have higher priority than existing requirements would be addressed in the next program and budget cycle.

QUESTIONS SUBMITTED BY MR. ROGERS

Mr. ROGERS. To my knowledge, the latest detailed GAO report that was issued specifically on Chemical Weapons Stockpile Emergency Preparedness in Alabama was issued in July 1996. At that time, the Army had allocated $46 million of Chemical Stockpile Emergency Preparedness Program (CSEPP) funding for Alabama. Today that figure has tripled. Are current summaries available that assess the current funding and status of CSEPP funding in Alabama? Additionally, the 1996 report included the impact of federal, state, and local management on Alabama's program. Have those findings changed significantly in 7 years, and if so, how?

Mr. HINTON. The amount of funding provided from the Army to the Federal Emergency Management Agency (FEMA) for enhancing emergency preparedness for tilecommunities surrounding the stockpile at Anniston, Alabama has grown subscuttially since the $46 million cited in our 1996 report.[4] In 2001, we reported the total amount of funding provided to the state of Alabama had increased to about $108 million—an increase of about $62 million from 1996.[5] However, Alabama was still lacking five critical items (over-pressurization, tone alert radios, coordinated plans, CSEPP staffing, and shelter in place kits) that would require additional funding. According to FEMA officials, about another $74 million will have been directly provided to Alabama by the end of fiscal year 2003—for a total of about $182 million for fiscal years 1989-2003.

In addition to the funding provided directly to Alabama, FEMA has provided funding for federally managed contracts on behalf of Alabama and to the U.S. Army Corps of Engineers for collective protection constuction projects in Alabama. According to FEMA, it has provided $63.6 million during fiscal years 1989-2003 for this indirect support which, when added to the direct funding mentioned above, totals about $246 million. This funding support has contributed to improvements in the status of community preparedness in Alabama. As of October 2003, Alabama was reporting that 7 of its 12 performance benchmarks are in compliance, and that the other 5, while not complete, are either on track for completion or only have minor problems. However, more funding than what is currently planned for Alabama may

---

[4] U.S. General Acconting Office, *Chemical Weapons Stockpile: Emergency Preparedness in Alabama Is Hampered by Management Weakness,* GAO/NSLAD-96-150 (Washington, D.C.: July 23, 1996).

[5] The $108 million reflects funding for fiscal years 1989 through 2000. See U.S. General Accounting Office, *Chemical Wealpons: FEMA and Army Must Be Proactive in Preparing States for Emergencies,* GOA-01-850 (Washington, D.C.: Aug. 13, 2001)

be needed to fully correct these problems. As I mentioned in my statement to the Committee, while about $41 million in direct funding is allocated to Alabama for fiscal years 2004 and 2005, FEMA identified unfunded requirements of about $38 million over the Same period.

GAO's work performed since the mid-1990s related to CSEPP management issues have shown that, at the federal level, the Army and FEMA have made significant progress in correcting management weaknesses, although there have been setbacks along the way. Since 1996, our work has not specifically focused on management issues related to the Anniston site, however, information concerning Anniston was included in our body of CSEPP work. The following chronicles past GAO work related to management issues:

- In 1996, we reported that there was a lack of CSEPP progress in Alabama because of management weaknesses at the federal level and inadequate action by state and local agencies. More specifically, we reported that the Army was slow to achieve progress because: (1) management roles and responsibilities were fragmented and unclear, (2) planning guidance was incomplete and imprecise, (3) budget processes lacked teamwork, and (4) financial controls were ineffective. In addition to progress being hampered by management weaknesses at the federal level, we reported that some state and local actions had contributed to delays in implementing a number of critical projects, such as issuing tone radios and conducting demographic studies.
- In a 1997 follow-up review that assessed the status of CSEPP in all 10 states, we reported that while the Army and FEMA had acted to improve program management, the effectiveness of these actions had been limited by differences regarding their respective roles and responsibilities.[6]
- Subsequently, we reported in 2001 that the Army and FEMA had improved their joint management of the program by improving working relations with each other and more clearly defining roles and responsibilities. However, we reported that they had not been as successful in their working relations with some states and local communities. With regard to Alabama, we reported that unresolved issues remained involving overpressurization projects and coordinated plans. These issues resulted in gaps in emergency response capabilities and were unresolved due to a lack of agreement between the Army, FEMA, and local officials.
- Last, in 2003, we reported that frequent shifts in program leadership had contributed to confusion among participants in program areas, including CSEPP, contributing to delays in Alabama.[7] Moreover, we also reported that the Army and some key Alabama stakeholders were not able to satisfactorily resolve issues such as the adequacy of protection plans [overpressurization projects] prior to the planned agent operations start date at Anniston, delaying operations many months. FEMA officials are now confident that most of the past issues related to Alabama are largely resolved and that working relations between the Army, FEMA, and the states have improved.

In responding to these questions, we relied primarily on our past work. For additional information on our work on DOD's chemical demilitarization program,

---

QUESTIONS SUBMITTED BY MR. FLETCHER

Mr. FLETCHER. Directives have been issued by various authorities within DOD and DOA to accelerate the disposal of the U.S. Stockpile of Chemical Weapons since the attacks of 9/11.

Does the Department have the necessary funds available in FY04 and FY05 and in their budget requests for FY06 - FY10, to maintain an accelerated disposal schedule for the Blue Grass Army Depot (BGAD)?

Mr. WAKEFIELD. Yes, the Department currently has all of the necessary resources, appropriated and/or programmed, to maintain an accelerated disposal schedule for the Blue Grass Chemical AgentDestruction Pilot Plant. This is based on the concept approved by the Defense Acquisition Executive in February 2003. If any new requirements are substantiated to sustain an accelerated program at Blue Grass are validated, the Department will work to obtain the necessary resources.

---

[6] U.S. General Accounting Office, *Chemical Weapons Stockpile: Changes Needed in the Management of the Emergency Preparedness Program,* GAO/NSIAD-97-91 (Washington, D.C.: June 11, 1997).

[7] U.S. General Accounting Office, *Chemical Weapons: Sustained Leadership, Along with Key Strategic Management Tools, Is Needed to Guide DOD's Destruction Program,* GAO-03-1031 (Washington, D.C.: Sept. 5, 2003).

Mr. PARKER. The Program Manager for Assembled Chemical Weapons Alternatives has identified activities that would allow the Department to maintain an accelerated schedule for the BGAD. The Army and the Department of Defense will address these issues in the upcoming programming and budgeting cycle.

○

THE IMPACT OF THE DRUG TRADE ON BORDER SECURITY AND NATIONAL PARKS

HEARING

BEFORE THE

SUBCOMMITTEE ON CRIMINAL JUSTICE, DRUG POLICY AND HUMAN RESOURCES

OF THE

COMMITTEE ON GOVERNMENT REFORM

HOUSE OF REPRESENTATIVES

ONE HUNDRED EIGHTH CONGRESS

FIRST SESSION

MARCH 10, 2003

Serial No. 108–19

Printed for the use of the Committee on Government Reform

Available via the World Wide Web: http://www.gpo.gov/congress/house
http://www.house.gov/reform

U.S. GOVERNMENT PRINTING OFFICE

87–703 PDF WASHINGTON : 2003

For sale by the Superintendent of Documents, U.S. Government Printing Office
Internet: bookstore.gpo.gov Phone: toll free (866) 512–1800; DC area (202) 512–1800
Fax: (202) 512–2250 Mail: Stop SSOP, Washington, DC 20402–0001

COMMITTEE ON GOVERNMENT REFORM

TOM DAVIS, Virginia, *Chairman*

DAN BURTON, Indiana	HENRY A. WAXMAN, California
CHRISTOPHER SHAYS, Connecticut	TOM LANTOS, California
ILEANA ROS-LEHTINEN, Florida	MAJOR R. OWENS, New York
JOHN M. MCHUGH, New York	EDOLPHUS TOWNS, New York
JOHN L. MICA, Florida	PAUL E. KANJORSKI, Pennsylvania
MARK E. SOUDER, Indiana	CAROLYN B. MALONEY, New York
STEVEN C. LATOURETTE, Ohio	ELIJAH E. CUMMINGS, Maryland
DOUG OSE, California	DENNIS J. KUCINICH, Ohio
RON LEWIS, Kentucky	DANNY K. DAVIS, Illinois
JO ANN DAVIS, Virginia	JOHN F. TIERNEY, Massachusetts
TODD RUSSELL PLATTS, Pennsylvania	WM. LACY CLAY, Missouri
CHRIS CANNON, Utah	DIANE E. WATSON, California
ADAM H. PUTNAM, Florida	STEPHEN F. LYNCH, Massachusetts
EDWARD L. SCHROCK, Virginia	CHRIS VAN HOLLEN, Maryland
JOHN J. DUNCAN, JR., Tennessee	LINDA T. SANCHEZ, California
JOHN SULLIVAN, Oklahoma	C.A. "DUTCH" RUPPERSBERGER, Maryland
NATHAN DEAL, Georgia	ELEANOR HOLMES NORTON, District of Columbia
CANDICE S. MILLER, Michigan	JIM COOPER, Tennessee
TIM MURPHY, Pennsylvania	CHRIS BELL, Texas
MICHAEL R. TURNER, Ohio	
JOHN R. CARTER, Texas	
WILLIAM J. JANKLOW, South Dakota	BERNARD SANDERS, Vermont (Independent)
MARSHA BLACKBURN, Tennessee	

PETER SIRH, *Staff Director*
MELISSA WOJCIAK, *Deputy Staff Director*
RANDY KAPLAN, *Senior Counsel/Parliamentarian*
TERESA AUSTIN, *Chief Clerk*
PHILIP M. SCHILIRO, *Minority Staff Director*

SUBCOMMITTEE ON CRIMINAL JUSTICE, DRUG POLICY AND HUMAN RESOURCES

MARK E. SOUDER, Indiana, *Chairman*

NATHAN DEAL, Georgia	ELIJAH E. CUMMINGS, Maryland
JOHN M. MCHUGH, New York	DANNY K. DAVIS, Illinois
JOHN L. MICA, Florida	WM. LACY CLAY, Missouri
DOUG OSE, California	LINDA T. SANCHEZ, California
JO ANN DAVIS, Virginia	C.A. "DUTCH" RUPPERSBERGER, Maryland
EDWARD L. SCHROCK, Virginia	ELEANOR HOLMES NORTON, District of Columbia
JOHN R. CARTER, Texas	CHRIS BELL, Texas
MARSHA BLACKBURN, Tennessee	

EX OFFICIO

TOM DAVIS, Virginia	HENRY A. WAXMAN, California

CHRISTOPHER A. DONESA, *Staff Director and Chief Counsel*
NICOLAS COLEMAN, *Counsel*
NICOLE GARRETT, *Clerk*